Contents

Preface   ix
Acknowledgments   xi
Introduction   xiii

Autobiographical Flights

Values from a Chicago Upbringing   3
Growing Up in the Phage Group   7
Minds That Live for Science   17
Early Speculations and Facts about RNA Templates   23
Bragg's Foreword to *The Double Helix*   33
Biographies: Luria, Hershey, and Pauling   37

Recombinant DNA Controversies

In Further Defense of DNA   49
Standing Up for Recombinant DNA   61
The Nobelist Versus the Film Star   71
The DNA Biochemical Canard   75

Ethos of Science

Moving Toward the Clonal Man: Is This What We Want?   83
The Dissemination of Unpublished Information   91
Science and the American Scene   105

The Necessity for Some Academic Aloofness   109
Striving for Excellence   117
Succeeding in Science: Some Rules of Thumb   123

War on Cancer

The Academic Community and Cancer Research   129
Maintaining High-Quality Cancer Research in a Zero-Sum Era   139
The Science for Beating Down Cancer   147

Societal Implications of the Human Genome Project

Moving on to Human DNA   163
Ethical Implications of the Human Genome Project   169
Genes and Politics   179
Five Days in Berlin   209
Good Gene, Bad Gene: What Is the Right Way to Fight the
   Tragedy of Genetic Disease?   223
Viewpoint: All for the Good—Why Genetic Engineering Must
   Soldier On   227

Afterword: Envoi—DNA, Peace, and Laughter   231

Name Index   239
Subject Index   245

To my sons
Rufus and Duncan

Introduction

Francis Crick—the other half of the famous partnership that lit the fuse and ignited the grand new enterprise of molecular biology—recorded in his autobiography (*What Mad Pursuit*) that at the time of their great discovery his friend Jim Watson was generally regarded as too bright to be sound. There is an echo here of a sage observation, buried in Max Beerbohm's fable of Oxford life, *Zuleika Dobson:* "The dullard's envy of brilliant men is always assuaged by the suspicion that they will come to a bad end." But Watson and Crick have in their different ways continued to exercise a pervasive influence on the course of science, as this collection of Watson's speeches, depositions, recollections, and ruminations makes clear.

Watson, as he relates in one of these pieces, did not quite qualify as a child prodigy. His early (and enduring) passion for ornithology propelled him into a radio quiz program in Chicago, but he had trouble with his maths, and although he got into the University of Chicago at fifteen (thanks to the liberal policy of its great President, Robert M. Hutchins), his high-school record did not betray outstanding talents. But by the time he entered graduate school his mentors, Salvador Luria and Max Delbrück, both themselves future Nobel Laureates, evidently discerned that the spindly nineteen-year-old was something out of the ordinary. François Jacob, in his absorbing memoir, *The Statue Within*, has given a vivid description of the young Watson, whom he found to be "an amazing character. Tall, gawky, scraggly, he had an inimitable style. Inimitable in his dress: shirttails flying, knees in the air, socks down around his ankles. Inimitable in his bewildered manner, his mannerisms .... A surprising mixture of awkwardness and shrewdness. Of childishness in the things of life and maturity in those of science." It is little wonder that Sir Lawrence Bragg, the Cambridge mandarin who found even the ebullient Crick hard to tolerate at times, did not know what to make of this strange visitor from Mars. There were others on whom Watson's brash and precocious self-assurance made a less than endearing impression, and he has told in his book, *The Double Helix*, how he fled before the wrath of an incensed Rosalind Franklin. Yet when I asked Maurice Wilkins, who shared the Nobel Prize for DNA with Watson and Crick, for his impression of Watson in 1952, he replied that he had found him very charming.

The great discovery that linked the names of Watson and Crick—one of the intellectual watersheds of the twentieth century—came when Watson was twenty-five, not as the culmination of a lifetime's toil. Watson did not find the burden insupportable; he has written that his pleasure in the achievement satisfied him for at most a month. Science can be an unforgiving profession. Its ceaseless exigencies are nicely encapsulated by a *New Yorker* cartoon, showing a Neanderthal of sour mien, sunk in thought and watched by two of his fellows: "So he invented fire and the wheel," the caption runs, "but what has he done since?" Watson, however, did not repine and, along with a band of fellow-frontiersmen, mainly in the two Cambridges, in California and in Paris, he helped the refulgent new discipline take shape.

What was it then about the famous double helix that so fired the imagination of the brightest minds of the time and drew even leading theoretical physicists (a breed of men accustomed to equating biology with stamp collecting), such as George Gamow and Leo Szilard, into its slipstream? There had been since 1944 compelling evidence that DNA was the genetic material that carried the blueprint of the organism—the homunculus in the sperm and the egg. But how all this information could be stored in a polymer—a string of linked chemical units (the nucleotides) of only four kinds—was a question that seemed to have no answer. Nor was it clear how the information could be replicated and passed to the offspring when cells divided. Watson and Crick felt (as did Maurice Wilkins and Rosalind Franklin in London, and Linus Pauling, the leading structural chemist of the day, in California) that if the structure of DNA could be determined, answers to these questions might emerge. Yet they were not wholly confident, and Watson wrote afterwards of the insidious fear that the structure might no more disclose the function of the molecule than had that of collagen (the structural protein of skin, and to Watson the most boring of biological entities).

The DNA structure came as an epiphany, for it immediately revealed that the two strands (twined around each other in the double helix) were complementary, related, in other words, rather as the photographic positive is to the negative. The bases of the nucleotide units pair off, each A with a T, each C with a G, the two strands running in opposite directions, one up and one down. The paper by Watson and Crick in *Nature*, in which this model was unveiled, concluded with one of the most famous sentences in the scientific literature, the teasing understatement: "It has not escaped our notice that the specific pairing we have postulated immediately suggests a possible copying mechanism for the genetic material." Sydney Brenner, who was to become a leader of the biological revolution but was then still a graduate student in Oxford, has written that the day in April of 1953 when he traveled to Cambridge to see the DNA model and meet its two begetters was the most exciting of his life. "The double helix," he concludes, "was a revelatory experience; for me, everything fell into place and my future scientific life was decided there and then." Watson himself dates the true identification of DNA as the genetic essence to 1953, for though he and others may have believed it to be such from the evidence already at hand, the lineaments of the model gave it substance.

Not everyone, it should be said, understood so clearly the meaning of the structure and what it portended. Many of the luminaries of biology and biochemistry simply did not grasp the point; others were shocked by the seeming amateurishness of Watson and Crick's model-building efforts. Erwin Chargaff was a respected biochemist who had devoted years of meticulous research to the analysis of the base composition (the proportions of the four nucleotides) in DNA samples from many sources. The important outcome was that the composition varied widely from one organism to another, but that the proportions of A and T were always the same, and that of C the same as G. The meaning of this conundrum ("Chargaff's rule") was made instantly manifest by the structural model. Yet Chargaff, so far from welcoming this startling illumination of his observations, simply could not believe that any good might have come from the fumbling incursions of two interlopers into his territory; he had been outraged to discover they were even hazy about the chemical structures of the four bases. Chargaff, a master of the pungent witticism, later dismissed molecular biology as no more than biochemistry practiced without a license.

Fifteen years after the event, Watson set down the story of the discovery as he remembered it. Of all the books about science, *The Double Helix* is perhaps the most widely read ever (but not the most widely bought: that distinction belongs to Stephen Hawking's *A Brief History of Time*). It was named by the literary committee behind the Random House Modern Library as number seven of the hundred most important books of the century, but it gives a highly atypical picture of how science normally proceeds, which is by a laborious process of trial and error, of opportunities missed and experiments gone wrong or misconceived, and then (with luck) a slow dawning of an answer. Watson pictures his time in Cambridge as an exuberant race for the Nobel Prize. It is no secret (indeed Watson in one of the reminiscences collected here enlarges on the issue) that both Crick and Wilkins saw it rather differently, to the point, indeed, that they sought to prevent the book from being published at all. They succeeded in securing its rejection by Harvard University Press (thereby considerably impoverishing the university), and for a time a certain *froideur* entered into Watson's friendship with Francis Crick. But that Watson prefigured so clearly the surpassing importance of solving the DNA structure stands as a remarkable testimony to the clarity of his scientific vision.

Of course it was not only Crick and Watson who were occupying themselves with DNA, and there can be little doubt that the structure would have emerged from other directions within a year or two. The helicity of the molecule was in the event scarcely relevant to the developments that followed; it was the complementarity of the two strands that directed the course of genetics for the next decade and more. Had Watson stuck it out in Copenhagen, as the terms of his fellowship instructed him, the essential features of the genetic material would probably have emerged little by little over the months and years. The pairing of A with T and G with C, adumbrated by Chargaff's ratios, could have been divined from chemical evidence (and almost had been when the British biochemist, J.M. Gulland, lost his life in a commuter train crash). Physical chemists, notably Paul Doty and his colleagues at Harvard, were on the verge of showing that DNA could be split into two by physical manipulation and must therefore be two-stranded. But then there would have been no comparable theatrical coup, no sudden illumination of the biological landscape that needed both the allure of the double helix (so striking a symbol that it was purloined for one of his paintings by Salvador Dali) and the implications of reciprocity between the strands. The whole was greater than the sum of its parts, and it was the manner as much as the substance of the discovery that made the impact and at once implanted a Nobel Prize in the womb of time. As the venerated biologist of an earlier era, the comte de Buffon, put it, *le style c'est l'homme même.*

But this is conjecture and has been questioned by no less an authority than Francis Crick himself: It was not he and Watson who made DNA, he tells us, but DNA which made them. What remains undisputed is that the DNA structure initiated an intellectual revolution that has given us answers to questions that have exercised the human mind since the dawn of reason. It has brought about cataclysmic advances in medicine and agriculture and spawned the biotechnology industry. The elucidation of the DNA structure was complemented by the work of Fred Sanger in Cambridge on proteins. Proteins also consist of long chains, the chemical units of which are amino acids, twenty in number. It had been commonly supposed that they were randomly assimilated into the protein when it was synthesized and were not arranged in any preordained sequence. But Sanger showed that the sequence (first determined for insulin) was in reality unique to a given protein. Crick and his friends at once perceived that there must be a correspondence of some kind between the sequence of bases in the DNA of a gene and the amino acid sequence of the protein to which it gave rise. This *aperçu* started the hunt for the genetic code, perhaps the most exciting episode in the early history of molecular biology. The discovery that sickle-cell disease, a life-threatening form of anemia, was caused by replacement of a single amino acid by another in the oxygen-carrying protein, hemoglobin, introduced the concept of a point mutation—a chance event of a kind that we now know to underlie the process of evolution, as well as of disease.

Other questions tumbled in torrents from the DNA structure: How did the separation and copying of the strands occur? What was the nature of the genetic code? To designate twenty amino acids, the code word had to comprise a run of not less than three letters (nucleotides), because the four kinds of nucleotides in the chain could give rise to only sixteen sequences of two, but there were sixty-four possible triplets. By what manner of mechanism was the information encrypted in the DNA sequence imposed on that of a protein chain? What was the role of the mysterious DNA-like polymer, RNA, which was metabolized so rapidly in active cells? Crick and Watson, joined by a growing (though for many years still select) band of mainly young biologists, chemists, and physicists, found answers in an astonishingly short time. (The story has been most vividly and perceptively told by Horace Judson in his book, *The Eighth Day of Creation*.) The field, in its early heady pioneering days, was characterized by a sense of collegiality, and, above all, candor ruled: Every proposition, each result was subjected to merciless criticism, and no feelings were spared. The milieu was reminiscent of the golden age of physics in the early part of the century: When the young Wolfgang Pauli prefaced his contribution to a discussion with the words, "you know, what Einstein just said is not so stupid," he was not thrown down the stairs.

The kitchen was too hot for some, who were intimidated by the ruthless intelligence of Crick and the other pacesetters. François Jacob relates that when Watson had shown his opinion of the talks at a small conference by ostentatiously reading a newspaper, the entire audience, when his own turn came to speak, drew newspapers from their pockets, as one man, and began to read.

While Francis Crick continued to change the face of science and is still at work in his eighties, Jim Watson, after many productive years as a professor at Harvard, where he nurtured a succession of brilliant and precocious *protégés*, took on the unlikely role of statesman. He became director of the decaying Cold Spring Harbor Laboratory on Long Island and converted it in the space of a few years into one of the world's preeminent centers of molecular genetics and cancer research.

Until perhaps twenty years ago, cancer was a field generally held in low regard. Funding was good, often profligate, but when so little was understood about how normal cells worked, there were poor hopes of formulating questions about why they went wrong. Moreover, the mandarins who dispensed the boodle were reluctant to allocate a place in their scheme of things to viruses, of which there was by then some understanding. When John Bittner in the fifties discovered that breast cancer in mice was caused by a virus he did not dare—for fear of being stigmatized as a crank—to call it by its proper name, and referred to it instead as a carcinogenic "factor." A report by the United States Surgeon-General had already excoriated all talk of cancer viruses. Peter Medawar's famous adage states that, if politics is the art of the possible, science is the art of the soluble; in the era before molecular biology the cancer problem was simply not soluble and was therefore profoundly unattractive to the best biologists. When Watson decided that its time had come, he proselytized vociferously for a program at Cold Spring Harbor and throughout the country, and the consequences in the laboratory, if not yet in the clinic, have been tumultuous.

Watson was also one of those who brought the human genome project—the decoding of our full complement of DNA (the genome), with its 100,000 or so genes, and much else still unrevealed—into being, not without some impassioned opposition. The carefree dawn, when a small fraternity of friends posed and often answered questions about the nature of life, lingers only in nostalgic recollection. Molecular biology has become Big Science. Its costs are rising inexorably to meet those of particle physics. It is evolving into an undertaking of big teams, fierce jostling for funds, secrecy, patents, and an uneasy relation to the huge industrial interests that have fed off its discoveries. As Watson's dissertations on the subject show, his science needs its statesmen as never before.

Watson is no stranger to controversy, and he has taken on the enemies of the calling that he represents with evident relish. There have been enemies within the gates, as well as outside. In the community of biologists voices were raised, protesting (with some justice) that molecular genetics was parasitizing the rest of biology. Some discerned unacceptable hazards in the new biology, and there were even those who saw in it little of intellectual or practical virtue. As late as 1977, when cloning and the methods of introducing foreign genes into bacteria for large-scale preparation of proteins were already well advanced, Sir Ernst Chain, a Nobel Laureate and one of the elders of the biochemical establishment, was able to write that such prospects belonged to the realm of science fiction. In just the same way, one of the leaders of the American war machine, Admiral Leahy, in June of 1945 denounced the Manhattan Project as "the biggest fool thing we have ever done. The bomb will never go off, and I speak as an expert on explosives."

These debates, at least, have now been won. Many biologists believe that they now face a new and more formidable enemy; the neo-Luddites (as their adversaries see them) recognize the power that the methods of molecular biology have delivered into the hands of today's scientists; they see moral dilemmas and physical threats all around them, ranging from Dolly the sheep to genetically modified crops and the prospects of finding our most intimate genetic traits exposed to bureaucratic inspection. Watson is not much troubled by their arguments. He is at heart an optimist, who believes that the risks are far outweighed by the benefits for prosperity and especially health that future generations will enjoy. Whether or not you agree with his reasons and embrace his conclusions, you will find in his writings the most lucid and authoritative statements for the defense. You will also find here a collection of reflections and reminiscences that give the authentic flavor of a period of momentous achievement and change, and of irreverence and boundless confidence.

Walter Gratzer

Autobiographical Flights

Values from a Chicago Upbringing

1993

Emotionally and intellectually I have been more formed by Chicago than by experiences anywhere else. Outsiders often sneer at Chicago, but I know otherwise. There, in 1928, I was born into a family with three paramount values. One was the importance of books and the belief that knowledge would liberate mankind from superstition, which for my father, brought up to be an Episcopalian, meant religion. The second value was birds. From his adolescence, my father was addicted to observing birds, and later I happily joined him knowing that it liberated me from the Sunday services of my mother's church. By the time I entered South Shore High School, I also was obsessive about trying to find rare birds in Jackson Park, around Wolf Lake, or out in the Indiana sand dunes at Tremont. It gave romance to my life and more than compensated for being even shorter than my finally 5′2″ sister, Betty, two years my junior. Our third family value was the nobility of the Democratic Party, led by my first real hero, Franklin Delano Roosevelt. Then if you had a family car, you could afford to be a Republican, but if you had been knocked down by the Depression, common sense made you a Democrat.

Beginning when I was about 12, my father and I every Friday evening made the mile-long walk to the library on 73rd Street to browse among its stacks and invariably bringing several home to digest during the following week. Our house was also filled with books, the more recent of which came from the Book of the Month Club, but most of which came from the used book shops in the Loop or Hyde Park. Dad worshiped persons of reason and took particular pleasure in reading the thoughts of the great philosophers. His library also had the occasional book on science, and it was those, not the books on philosophy, which I pored over when the weather

3

was too unpleasant for bird watching. Learning about evolution particularly caught my fancy, with Darwin's theory of natural selection providing
a rational way to think about the diverse forms of life that first excited me
through trips to the Field Museum.

Even before I entered the University of Chicago in 1943, I had begun
to daydream about being a scientist, though I had to wonder whether I was
bright enough to enter this world filled with geniuses. All too clearly I had
not entered the University of Chicago at 15 because of my high IQ. I was
far from the child genius that say Wally Gilbert, several years later, grew up
to be in Washington. My premature departure from South Shore High
School instead reflected the fact that Robert Hutchins, still the almost boy
president of the University of Chicago, considered American high schools
disasters that never could be reformed. And instead of wasting money failing to improve them, he had the simple solution of getting kids into college two years earlier. That I was one of the first entrants into Hutchins'
Four Year College owed much to my Southside-Irish raised mother, who
had gone to the nearby University of Chicago. It was she who saw that I
filled out the application form for a tuition scholarship which later let me
attend college, initially needing only from my family the two three-cent
fares for the daily streetcar ride of some 30 minutes.

My first two years at the U of C were superficially not very successful,
with my grades (largely B's) continuing to expose to all my non-genius
qualities. But they prepared me for the future by instilling upon me three
new values. The first was to focus on original sources instead of textbooks
— read the great books themselves, not the interpretations of others. The
second value was the importance of theory. Of course, you have to know
some facts, but much more important is how to put them together in some
rational scheme. And thirdly, you had to concentrate on learning how to
think as opposed to improving memorization skills. Initially, to my annoyance, the big comprehensive exams that gave us our grades for the entire
year often seemed to bear no relation to what you learned in your lectures.
With time I realized that I did not have to take notes but instead could concentrate on whether the lecturer's words actually made sense. In retrospect, I now realize I was acquiring the mental habits which later made me
acceptable first to Luria and Delbrück and later to Francis Crick.

And Chicago being then the Second City and the University of Chicago not as old as Harvard, I saw no reason to treat authority with much rev-

erence. You were never held back by manners, and crap was best called crap. Offending somebody was always preferable to avoiding the truth, though such bluntness did not make me a social success with most of my classmates. It was lucky that for much of my college life, I was still too short to see the need to effectively move outside the security of my family home. But being honest about what is bad and false leads nowhere, unless you hold equally strong values about what is good. To escape from the false leads of the present, you have to have good advice as to where science should go. Happily, my first scientific hero was the famed U of C geneticist, Sewell Wright, whose course on physiological genetics I audited in the spring of 1946.

The gene had suddenly come to the forefront of my attention several months before through reading "What Is Life?" by the Austrian theoretical physicist, Erwin Schrodinger. Soon after its publication in the States, I spotted this slim book in the Biology Library, and upon reading it was never the same. The gene being the essence of life was clearly a more important objective than how birds migrate, the scientific topic that previously I could not learn enough about. So I considered myself very fortunate that as a Zoology major, I had the opportunity three times a week to listen to one of the world's best geneticists, whose interests ranged from how genes work to the mathematical foundations of evolutionary genetics. Even though Sewell Wright was not an inspired teacher, using notes written on 3 x 5" file cards, I never wanted his lectures to end. By the term's end, I had made the decision to have the gene as my life's principal objective.

So I now feel very warm toward the City of Chicago and its great university on the Midway, which gave me not only the intellectual goal that has so dominated the remainder of my life but also the intellectual obstacle course that its great books demanded I jump over. Of course, I would never have come to the Double Helix without the subsequent testing of many other intellectual waters. But it was Chicago that first taught me that I had to be different from others if later I was to succeed. But my teachers were not there to say at what cost—to both me and those you want to believe in you.

Growing Up in the Phage Group

1966

As an undergraduate at Chicago,[1] I had already decided to go into genetics even though my formal training in it was negligible, with most of my course work reflecting a boyhood interest in natural history. My obvious choice for graduate school was Caltech, because I was told its Biology Division was loaded with good geneticists. They, however, did not want me, nor did Harvard, to which I had applied without considering what I might find. Harvard's lack of interest in me was particularly fortunate, because if I had gone there I would have found no one excited by the gene and so might have been tempted to go back into natural history. Fortunately my advisor at Chicago, the human geneticist Strandskov, also had me apply to Indiana University in Bloomington, emphasizing that H.J. Muller was there as well as several very good younger geneticists (Sonneborn and Luria). To my relief, Indiana took a chance with me, offering a $900 fellowship for the coming 1947–1948 academic year. Characteristically, Fernandus Payne, then dean of its graduate school, wanted to make sure that I knew what I was getting into. He appended a postscript to the fellowship offer saying that if I wanted to continue my interest in birds I should go elsewhere.

---

[1]At the age of nineteen in 1947, Watson graduated from the University of Chicago and prepared to enter graduate school.

During my first days at Indiana, it seemed natural that I should work with Muller,[2] but I soon saw that Drosophila's better days were over and that many of the best younger geneticists, among them Sonneborn and Luria, worked with microorganisms. The choice among the various research groups was not obvious at first, because the graduate student gossip reflected unqualified praise, if not worship, of Sonneborn. In contrast, many students were afraid of Luria, who had the reputation of being arrogant toward people who were wrong. Almost from Luria's first lecture, however, I found myself much more interested in his phages than in the Paramecia of Sonneborn. Also, as the fall term wore on I saw no evidence of the rumored inconsiderateness toward dimwits. Thus with no real reservations (except for occasional fear that I was not bright enough to move in his circle) I asked Luria if I could do research under his direction in the spring term. He promptly said yes and gave me the task of looking to see whether phages inactivated by X rays gave any multiplicity reactivation.

The only other scientist in Luria's lab then was Renato Dulbecco, who six months previously had come from Italy to join in the experiments on the multiplicity reactivation of UV-killed phages. I was given a desk next to Dulbecco's and, when he was not doing experiments, often worked on his lab bench. Most of Luria's and Dulbecco's conversation was in Italian, and I might have felt somewhat isolated had it not been for the fact that my first experiments gave a slightly positive result. Usually, Luria never let even a few hours pass between the counting of my plaques and his know-

[2]Hermann J. Muller was one of the leading geneticists of his day. A convinced communist, he had in 1933 taken up an appointment in Soviet Russia. His interest in eugenics, which he hoped to develop in Moscow, had however alienated him from official party doctrine. His book, *Out of the Night: A Biologist's View of the Future*, included a discussion of large-scale artificial insemination with the sperm of highly intelligent donors. It was Muller's additional misfortune to fall foul of Lysenko. The rise to official favor of this demagogic charlatan culminated in the almost total suppression of genetic research in the Soviet Union. Muller disobeyed the order to eschew all mention of human genetics at a fateful conference, held in Moscow in 1936, and was forced to flee for his life. He left the country with a transport of "volunteers" bound for the Civil War in Spain. Most members of his laboratory were arrested, and several of them (including the secretary who had translated his genetics textbook into Russian) were shot.

In 1946 he received the Nobel Prize in Physiology or Medicine for his 1926 discovery that X-rays cause mutations. He had then just moved to Indiana University, the sole American institution willing to chance bringing to its faculty his contentious political past.

ing the answer. Also, Dulbecco's family had not yet come from Italy, and we would occasionally eat together at the Indiana Union. During one Sunday lunch, I remember asking him whether Luria's figure of 25 T2 genes should not tell us the approximate size of the gene since the molecular weight of T2 could be guessed from electron micrographs. Dulbecco, however, did not seem interested, perhaps because he already suspected that multiplicity reactivation of UV-killed phage was more messy than Luria's pretty subunit theory proposed. Then there was also the fact that despite Avery, McCarthy, and MacLeod, we were not at all sure that only the phage DNA carried genetic specificity.[3]

Some weeks later in Luria's flat, I first saw Max Delbrück, who had briefly stopped over in Bloomington for a day. His visit excited me, for the prominent role of his ideas in *What is Life?*[4] made him a legendary figure in my mind. My decision to work under Luria had, in fact, been made so quickly because I knew that he and Delbrück had done phage experiments together and were close friends. Almost from Delbrück's first sentence, I knew I was not going to be disappointed. He did not beat around the bush and the intent of his words was always clear. But even more important to me was his youthful appearance and spirit. This surprised me, for without thinking I assumed that a German with his reputation must already be balding and overweight.

Then, as on many subsequent occasions, Delbrück talked about Bohr and his belief that a complementarity principle, perhaps like that needed for understanding quantum mechanics, would be the key to the real understanding of biology. Luria's views were less firm, but there was no doubt that on most days he too felt that the gene would not be simple and that high-powered brains, like Delbrück's or that of the even more legendary Szilard, might be needed to formulate the new laws of physics (chemistry?)

---

[3]Oswald Avery and his colleagues, working at the Rockefeller Institute in New York, showed in 1944 that DNA could be transferred from bacteria of one strain to those of another and that it brought with it the genetic attributes of the donor strain. This is generally held to mark the identification of DNA as the genetic material, but doubts lingered regarding some active impurity (probably a protein).

[4]*What is Life?* (Cambridge University Press, 1994) was an influential (though in retrospect largely wrong) book by the famous Austrian theoretical physicist Erwin Schrödinger, which persuaded many physical scientists after World War II that genetics would be a rewarding pursuit.

upon which the self-replication of the gene was based. So, sometimes I worried that my inability to think mathematically might mean I could never do anything important. But in the presence of Delbrück I hoped I might someday participate just a little in some great revelation.

I looked forward greatly to the forthcoming summer (1948) when Dulbecco and I would go with the Lurias to Cold Spring Harbor. Delbrück and his wife Manny were coming for the second half while, before they arrived, there was to be the phage course given by Mark Adams. No great conceptual advances, however, came out that summer. Nonetheless, morale was high even though Luria and Dulbecco sometimes worried whether they had multiplicity reactivation all wrong. Delbrück remained confident, however, that multiplicity reactivation was the key breakthrough which soon should tell us what was what. His attention, however, was then often directed toward convincing us that an argument of alternate steady states would explain Sonneborn's data on antigenic transformations in Paramecia. The idea of cytoplasmic hereditary determinants did not appeal at all to Delbrück and he hoped we would all join together to try to bury as many of them as possible.

As the summer passed on I liked Cold Spring Harbor more and more, both for its intrinsic beauty and for the honest ways in which good and bad science got sorted out. On Thursday evenings general lectures were given in Blackford Hall by the summer visitors and generally everyone went, except for Luria who boycotted talks on extra-sensory perception by Richard Roberts and on the correlation of human body shapes with disease and personality by W. Sheldon. On those evenings, as on all others, Ernst Caspari opened and closed the talks, and we marveled at his ability to thank the speakers for their "most interesting presentations."

Most evenings we would stand in front of Blackford Hall or Hooper House hoping for some excitement, sometimes joking whether we would see Demerec going into an unused room to turn off an unnecessary light. Many times, when it became obvious that nothing unusual would happen, we would go into the village to drink beer at Neptune's Cave. On other evenings, we played baseball next to Barbara McClintock's cornfield, into which the ball all too often went.

There was also the fair possibility that we could catch Seymour Cohen and Luria each informing the other that his experiments were not only over-interpreted but off the mainstream to genuine progress. Though

Cohen was spending the summer doing experiments with Doermann, a sharp gap existed between Cohen and the phage group led by Luria and Delbrück. Cohen wanted biochemistry to explain genes, while Luria and Delbrück opted for a combination of genetics and physics.

Cohen was not, however, the only biochemist about. David Shemin spent most of the summer living in Williams House while Leonor Michealis stayed for several weeks, despite his wife's complaints about the run-down condition of their apartment and of Demerec's failure to replace a toilet seat containing a large crack. When August began the Lurias went home to Bloomington because Zella Luria would soon have a child. This left Dulbecco and me even more free to swim at the sand spit or to canoe out into the harbor often in search of clams or mussels.

By the time we were back in Bloomington, all of us were again ready for serious experiments. Soon Dulbecco found photoreactivation of UV inactivated phage,[5] thereby explaining why the plaque counts in multiplicity reactivation experiments were often annoyingly inconsistent. This discovery did not seem a pure blessing, however, for it immediately threw into doubt all previous quantitative interpretation of multiplicity reactivation. Thus much of the work of the previous eighteen months had to be repeated, in both the light and dark. When this was finally done, it became clear that multiplicity reactivation was, by itself, not going to yield simple answers about the genetic organization of phage. As a corollary, my study of X-ray-inactivated phage also was much less likely to yield anything very valuable. By then, however, I had begun to study the indirect as well as the direct effect of X rays, and the complexity of the inactivation curves initially kept me from worrying whether they would be very significant.

That fall I had my first extended view of Szilard, when Luria, Dulbecco and I drove up to Chicago to see him and Novick. There I first realized that most conversations with Szilard occurred during meals, which seemingly consumed half of his time awake. Briefly I tried to tell him what I was up to, but soon I was crushed by his remark that I did not know how to speak clearly. Even more to the point was that Szilard did not like to learn new facts unless they were important or might lead to something important. Szilard and Novick later came to Bloomington for a small

[5]Irradiation with X rays or ultraviolet light had been shown to cause genetic changes (mutations). These were later demonstrated to result from chemical alterations in DNA.

phage meeting in the spring of 1949. Hershey, Doermann, Weigle, Putnam, Kozloff, and I were also there. Doermann had just done his genetic crosses using premature lysates and guessed that the percentage of recombinants was approximately constant throughout the latent period. Stent and Wollman described how they thought T4 interacted with tryptophan. To me the most memorable aspect of the meeting, however, was the unplanned comic performances of Szilard and Novick. Neither understood the other's description of their phenotypic mixing experiments and they were constantly interrupting each other, hoping to make the matter clear to everyone else. A day later, Delbrück, Luria, Dulbecco, and I drove to Oak Ridge for its spring meeting where Delbrück coined the phrase "The Principle of Limited Sloppiness" in explaining how Kelner and Dulbecco came upon photoreactivation.

The following summer Manny Delbrück was expecting a child and most of the phage group congregated in Pasadena instead of Cold Spring Harbor. Several times each week, there occurred seminars dominated by Delbrück's insistence that the results logically fit into some form of pretty hypothesis. There were also innumerable camping trips occupying two to four days, long weekends often led by Carleton Gajdusek whose need for only two or three hours sleep a night allowed him to spend five or six days each week in the wilderness while maintaining the pretense that he was interested in the world between John Kirkwood and Delbrück. Because Gunther Stent shared a house in the San Gabriel foothills with Jack Dunitz, Pauling's postdoctoral student, there were frequent social contacts with the younger students who worked for Linus Pauling but, on the whole, I never got the impression that the phage group thought that Pauling's world and theirs would soon have anything in common. Occasionally, I would see Pauling drive up in his Riley, and I felt very good when once he spontaneously smiled at me in the Faculty Club.

Most of the scientific arguments that summer were kinetic either about how tryptophan affected T4 adsorption or attempting to make sense of photoreactivation. Sometimes the genetic results of Hershey came into the picture, but only Doermann seemed tempted to do more along that line. My experiments on X-ray phage had progressed to the point where I knew I had a thesis, and so in Pasadena I played about a little with formedaldehyde-inactivated phages. Delbrück, like everyone else, was only mildly interested in my results but told me that I was lucky that I had not found anything as exciting as Dulbecco had, thereby being trapped into a rat race where people wanted you to solve everything immediately. If that had happened, he felt I would lose in the long term by not having the time either to think or to learn what other people were doing. I of course wanted something important to emerge from the masses of survival curves that filled several thick loose-leaf notebooks. Late in the evenings, I would try to imagine pretty hypotheses that tied all of radiation biology together, but so much special pleading was necessary that I almost never tried to explain them to Luria, much less to Delbrück.

In the early fall the question came up where I should go once I got my Ph.D. Europe seemed the natural place since, in the Luria–Delbrück circle, the constant reference to their early lives left me with the unmistakable feeling that Europe's slower-paced traditions were more conducive to the production of first-rate ideas. I was thus urged to go to Herman Kalckar's lab in Copenhagen, because in 1946 he had taken the phage course and now professed to want to study phage reproduction. Though Kalckar was admittedly a biochemist, through his brother he knew Bohr and thus should be receptive to the need of high-powered theoretical reasoning. Even better, Kalckar's interest in nucleotide chemistry should immediately be applicable to the collection of nucleotides in DNA. The decision was finally settled in mid-December, when by accident Kalckar and Delbrück both were in Chicago on a weekend when Szilard had got the midwestern phage people together for a small meeting. Kalckar seemed excited about the possibility of using some ^{14}C-labeled adenine, which had just been synthesized in Copenhagen, to study phage reproduction, and he gave the impression of very much wanting phage people to come to his lab.

The midwestern phage meetings were then being held almost every month in Chicago, thanks to a small grant to Szilard from the Rockefeller Foundation, which covered some of the travel expenses and all of the food bills. Lederberg also began to come, adding a new vocal dimension, because he could give nonstop 3- to 4-hour orations without making a dent in the experiments he thought we should know about. By then, he and his wife had found phage λ in *E. coli* strain K12, but perhaps because of Delbrück's dislike of the possibility of lysogeny, I paid little attention to the discovery. Instead I conserved my brain for the facts about the somewhat messy partial diploids. My guess is that no one left the meetings remembering more than a small fraction of the ingenious alternative explanations that Lederberg dreamed up to explain the increasing number of paradoxes arising from his experiments.

In the spring of 1950 Luria went back to the problem of the distribution of spontaneous mutations within single bursts of infected cells, hoping he would find out whether or not the genetic material replicated exponentially. I spent a month on the first version of my thesis, but Luria did not like it and took it home for rewriting. This left me little more to do and not surprisingly the thesis was accepted without fuss at my Ph.D. exam in late May. Then I went out to Pasadena for a month, flying back East to spend a final six weeks in Cold Spring Harbor before the boat would take me to Europe. For a brief while, I was afraid that the outbreak of the Korean War might keep me from sailing, but without hesitation my draft board gave me permission to leave the country.

Practical jokes dominated the mood during the late summer in Cold Spring Harbor, culminating in an evening when Gordon Lark, Victor Bruce and his sister, Manny Delbrück, and I let the air out of the tires of several friends' cars parked before Neptune's Cave. Afterwards, buckets of water were poured over our beds. On another evening, Visconti interrupted a staid Demerec social evening with an attack with a toy machine gun.

The growing number of phage people became noticeable at the phage meeting in late August. Some thirty people came, I being most affected by the talk of Kozloff and Putnam on their failure to observe 100% transfer of parental phage P32 to the progeny particles. Instead they believed that only somewhere between 20% and 40% of the parental label was transferred. While there existed loopholes in their experiments, the possibility was raised that perhaps both genetic and nongenetic phage DNA existed and that only the genetic portion was transferred. That prompted Seymour Cohen to predict that a second generation of growth might yield 100% transfer.

These ideas I followed up as soon as I got to Europe. Gunther Stent had also chosen Copenhagen, and so Kalckar was faced with two phage people far less interested in biochemistry than he had been led to expect. At the same time, when we could follow Kalckar's words, it was apparent that he was not fixed on the problem of gene replication and seemed happier talking about nucleoside rearrangements. Luckily Kalckar's close friend, Ole Maaløe, had been bitten with the phage bug, and without ever formally acknowledging the arrangement, Stent and I began working with Maaløe in his lab at the State Serum Institut. Maaløe liked the idea of the second-generation experiment, and we began making labeled phage. After a few failures we obtained the clean-cut, but then disappointing, result that the transfer in the second generation was the same as that in the first generation. The data were quickly written up and dispatched in early February (1951) to Delbrück for his approval and possible transmission to the Proceedings of the National Academy. My turgid style was quickly rejected by Delbrück, who completely rewrote the introduction and discussion sections and then sent it on.

By then I knew that Maaløe wanted to go to Caltech the following autumn, and so I had to find a place for the next year. I thus wrote to Luria of my dilemma, indicating a preference for England and mentioning Bawden and Pirie, neither of whom I had met. In Luria's reply he took me to task for laziness, saying that I should use my time to acquire the physics and chemistry necessary for a real breakthrough. Clearly my Copenhagen period was not developing the way Luria wanted it. Instead of learning anything new, Stent and I were merely transferring the phage group spirit to Denmark. The net result would be that I would end up doing routine phage work, and if that were to be the case it would make better sense for me to be in the United States.

Some months later, Luria responded more warmly when I suggested that I go to Perutz's group at the Cavendish Laboratory, to work on the structures of DNA and the plant viruses. Soon after my letter came, he met John Kendrew at Ann Arbor and set into motion the events that led me to Cambridge, Francis Crick, and the DNA structure.

Minds That Live for Science

⟨✕✕✕⟩

1987

I came to work at the MRC Unit for the Study of Structure of Biological Systems in September 1951. The unit then was four years old and occupied several rooms on the first floor of the Austin Wing of the Cavendish Laboratory of Cambridge University. Though Ernest Rutherford, the greatest experimental physicist ever to live, had been dead for less than 15 years, he was already a distant memory. Six grim years of the Second World War separated us from the nuclear physics of Rutherford's Cavendish days. This time the enemy was not only Hitler and the dreaded Wehrmacht, but also the skilled science and technology of the German people. The British triumph against such great odds owed much to the extraordinary use of its own scientific talents. The ending of the war thus released back to civilian life large numbers of already accomplished young men, exposed to sharp, nonemotional thinking and the need to aim for the best.

The then tiny unit, composed of Max Perutz, John Kendrew, Francis Crick, and Hugh Huxley, with Sir Lawrence Bragg, the Cavendish Professor, as its very involved patron, had as its objective the understanding of life at its deepest level, the molecular. By so doing, they hoped to transform biology from a morass of seemingly limitless and often boring facts into an intellectually satisfying discipline like physics or chemistry. The members of the unit worked and thought within a Cambridge populated by many others of great talent. Prominent among these stars of the future were the members of the biophysics-oriented Hardy Club that met in college rooms on the occasional Friday night. There I soon met Allen Hodgkin, Andrew Huxley, Victor Rothschild, Horace Barlow, Murdoch Mitchison, and Michael Swann. By this time the descendants of F. G. Hopkins's great biochemical days along Tennis Court Road were often more lost than inspired, but this

17

was not true of the protein chemists associated with A. C. Chibnall. There, in a small, hutlike lab, were Kenneth Bailey, Fred Sanger, and Rodney Porter. More difficult to assess was the charmingly long-haired possessor of a Rolls-Royce, Peter Mitchell, who was unconvinced that Francis Crick and I would succeed in hunting for templates instead of enzymes. We also saw much of Roy Markam and John Smith ("little Smith"), then supported by the more conventional plant virologist Kenneth Smith ("big Smith"), all located on the upper floor of the Molteno Institute. Then, due to the asthma of its director David Keilin, the Molteno was the only building in Cambridge with effective central heating. Even closer by was Alex Todd, the bigger-than-life-sized organic chemist from Glasgow who liked chemistry with a biological implication. After the double helix came out he congratulated Crick and me for our triumph as organic chemists. This highly diverse group of Cambridge academics focused on biology in a vastly more inspired way than any I had ever experienced at any American university. Except for their almost nonexistent genetics component, they were easily the intellectual equal of, if not better than, the collective best found all over the United States.

The professor of genetics then at Cambridge was R.A. Fisher, even by English standards an extraordinary, white-bearded eccentric, who left his wife and eight children in London when he acquired the Cambridge chair. He presided over an almost nonexistent department located out from the center of Cambridge on Story's Way to which, despite partial blindness, he daily cycled to and from his rooms in Caius College. His prestigious talents were those of the mathematician, not the maker of genetic crosses. Very unlike the even more clever J.B.S. Haldane, one of his very few intellectual equals as a population geneticist, Fisher had no deep interest in the nature of life. Instead he always gave the impression that mathematical truths were better than those achieved by experimentation, and near the end of his life he used specious arguments to contend that smoking and lung cancer were unconnected.[1]

---

[1]Fisher's work on the epidemiology of lung cancer was supported by the tobacco industry, a fact that caused much rancor when it became known. He argued that the disposition to lung cancer and the disposition to smoke tobacco might be linked. An analogy, drawn by a journalist, was how to interpret the supposed predominance of bald heads in the audiences at strip clubs: It is well-established that high levels of testosterone are causally linked to baldness, and so highly sexed men are more likely to be bald. The incorrect conclusion would be that watching ecdysiasts in action causes your hair to fall out. But Fisher's hypothesis was wrong: Smoking does cause lung cancer.

The status of genetics was even worse at Oxford, which had no chair in genetics at all. Only in 1953 did Oxford appoint as its professor of botany the clever but argumentative cytologist C.D. Darlington, who was dominated by what chromosomes look like under conventional microscopes and was never able to move beyond the 1930s. Then the brightest hope for genetics in Great Britain was to be Edinburgh. Its newly appointed professor of genetics, C.H. Waddington, who wrote the only first-class book in English on modern genetics in 1938, had been given the power and money to do something big. Unfortunately, his initial postwar efforts focused more on Marxism than on either Mendel or molecules. Originally trained as an embryologist, not as a geneticist, he was never satisfied with the dominant role of genes located in the nucleus and instead believed that non-DNA-determined "epigenetic" phenomena were the key to understanding the complexities of multicellular existence. Later Waddington retreated from experimental science, and as a "theoretical biologist" he organized meetings on the relations between the arts and sciences. Just after the double-helix model, I went up to Edinburgh to give a talk arranged by Michael Swann, the newly appointed professor of zoology. I stayed with Waddington at his early 19th-century Roman-styled villa and, knowing only of his splendid text, expected that he would be excited by the implications of the double helix. But to my dismay he seemed not to grasp what it meant, perhaps reflecting his wish for life to have unique biological explanations as opposed to being the consequences of the laws of chemistry.

My arrival in Cambridge and that of Sydney Brenner from South Africa some months later in Oxford thus added badly needed doses of genetic fresh air to the British biological world. We were each more excited about the possibilities of molecular genetics than any of our British peers, and by virtue of facts already assimilated in our brains, we were helping to fill an embarrassing vacuum. Unfortunately, Sydney went to an Oxford totally indifferent to what genes are and how they control the synthesis of proteins. Wanting to work on phages, Sydney placed himself under Sir Cyril Hinshelwood's guidance, not really believing that his supervisor, already a world-renowned physical chemist, could continue to dismiss the fluctuation test of Luria and Delbrück and would persist in his mad belief that chemical kinetics, not genes, made cellular existence possible.

At the same time I was the only individual in Cambridge who lived solely to understand how DNA functioned as the gene and who had first-hand practical experience in using bacterial viruses to get close to the self-replication of the gene. While I made repeated grasps at actually learning diffraction theory at the level of the Fourier transform, I never really needed to because Francis more than knew it, and besides we never had precise diffraction data to interpret. Also, I kept hoping that pure model-building would do the trick and that I could accomplish my goal without my total ignorance of physics ever being revealed.

From the start we hoped for some sudden chemical revelation that would lead to the correct structure, but we never anticipated that the answer would come so suddenly in one swoop and with such finality. As soon as the base pairs fell out,[2] we and everyone around us knew that we were onto something of immense importance. I saw the double helix as the culmination of almost a century of genetics, but for Francis it was to be the splendid beginning of a new life not only for him but for biology itself. Not less important for those who had supported us, our success was also the first major triumph of the unit and thus virtually ensured its long-term future. More slow to evolve was the realization that my return to the United States demanded the recruitment of someone to take my place as a geneticist and to provide a colleague for Francis, who Fred Sanger was known later to call "the geneman." But my replacement, Sydney Brenner, did not arrive for more than three years, in part because Francis himself was to spend a year in Brooklyn.

From Cambridge I went on to Caltech for two years, unsuccessfully trying to find the structure of RNA. Before I subsequently started teaching at Harvard in 1956, I returned to Cavendish for one last year to work with Francis and Don Caspar on the structure of small RNA viruses. Since then I have been back for two short sabbatical periods, and now return once or twice a year for even briefer visits. Each time I come up from London I approach Cambridge with the sense of coming home to a way of life that totally transformed my attitudes toward how to live. It left me (not unhappily) emotionally centered halfway between the East Coast and England and certainly never capable of the superficially unmannered life of California.

[2]Knowing exactly the shape of the four bases—the parts of the nucleotides that distinguish one from another—was critical for building a realistic model. Arranged correctly the bases from opposite strands of the double helix fit together like pieces of a jigsaw puzzle.

The fact that the LMB[3] has and still provides such extraordinary science produces in me both satisfaction and envy. The satisfaction comes eagerly from my role in creating the first of its now many supersuccesses. The envy comes from the fact that I am no longer part of what ranks as the most productive center for biology in the history of science. In its own way, its accomplishments at least equal the feats for physics by J.J. Thomson and Rutherford and their collaborators during an earlier Cavendish era. Many labs now do superbiology, but not one approaches what Max Perutz and John Kendrew started with Sir Lawrence Bragg's help.

In looking back to why the LMB has so led the pack, I must start with the extreme intelligence and scientific drive of its scientific leaders. There is no substitute for sharp, interactive brains. Going as a visiting scientist to the LMB to work was and still is an intellectual experience, sometimes fun, other times grueling, and seldom not tense in the long run. It is not the place for those weak in spirit to seek comfort with minds that no longer live for science. A second reason for its greatness is the high, if not heroic, level of so many of its goals. Indispensable for these lofty aspirations has been the long-term stable funding provided by the MRC. It has allowed problems to be attacked that could take 5 to 10 years to bring to fruition. A third ingredient for its success has been its organization into a series of research groups, many of whose members have tenure. This is a feature virtually never found in any American university lab, where more often than not the only person with tenure is the leader of the lab group. The LMB's way of doing science would have been risky, however, if there had not been a constant infusion of talented visiting postdocs, initially largely Americans, each of whom needed to achieve real success during the course of their two to three years at the unit and who therefore were often real workhorses of the LMB. And finally the siting of LMB within the extraordinary beauty of Cambridge, the most attractive site for science in Britain, if not the world, gave to the LMB the potential to attract very high-level research students.

The glorious current state of the LMB, which has had so much positive influence on the intellectual life of the whole world, has not been

---

[3]LMB, or the Laboratory of Molecular Biology, is what the small Medical Research Council in the Cavendish Laboratory eventually became. It now occupies a large building of its own.

achieved without its negative consequences. The combination of tenure of so many of its staff coupled with research conditions superior to virtually any elsewhere in Britain has meant that only a handful of its staff have moved on to create new centers of excellence at other British universities. This situation stands in sharp contrast with the great days of Rutherford, whose junior stars quickly took up chairs elsewhere. Contributing to this failure to help to develop a vigorous molecular biology tradition elsewhere in Britain has been the emphasis upon the training of foreign postdocs who come with external stipends. Graduate students who need to be paid with British monies have never played the major role that they do in the leading labs in the United States. Wally Gilbert and I, for example, at Harvard alone may have trained as research students more future leaders of molecular biology than has the LMB over its entire history. By not allotting money to support a vigorous program for the training of large numbers of clever research students, the LMB has not produced, and still is failing to produce, the new generation of competitively trained British scientists that will be needed all too soon to maintain the LMB and other major British labs in the forms we have so long admired.

Early Speculations and Facts about RNA Templates

1993

RNA first came alive to me during the fall of 1947 at Indiana University when I took Salvador Luria's course on viruses. Then I first learned that while the then-known phage, pox, and papilloma viruses contained DNA, it was totally lacking in several purified plant viruses as well as in the viruses that caused encephalitis, which instead contained RNA. Apparently a given virus had either RNA or DNA in contrast to cells which contained both. But whether it was the nucleic acid component that carried their genetic specificity was still unclear. Then most scientists wanted Avery, MacLeod, and McCarty's experiment on pneumococcal transformation by purified DNA to be extended to other life forms before jumping on the nucleic acid bandwagon. The spring 1952 report from Al Hershey, that the DNA component of phage T2 carried genetic specificity, immediately thrilled me, but I remember well the audience's indifference when in mid-April I read Hershey's letter to an Oxford meeting of The Society for General Microbiology.

When I arrived at Cambridge in the fall of 1951, I started taking seriously the work of Brachet and his collaborators in Brussels who emphasized the correlation between the RNA content and the protein synthesizing capacity of cells. Those cells making large amounts of protein possessed large numbers of virus-sized ribonucleoprotein particles, known initially as microsomal particles but since 1958 as ribosomes. Most importantly, these particles had been pinpointed as the actual sites of protein synthesis through just developed cell-free systems for protein synthesis. Here the key lab was that of Paul Zamecnik at Massachusetts General

Hospital. Equally important was Brachet and Chantrenne's demonstration that the nucleus and hence DNA had no direct participation in protein synthesis. To show this they cut the giant algae Acetabularia in two halves and observed that the half without any nucleus could maintain almost normal protein synthesis for more than a month. Yet from the one gene-one enzyme (protein) results of Beadle and Tatum, the ultimate source of the genetic information that specifies the amino acid sequences of proteins had to be the genes found in the nucleus. I thus postulated a two-stage scheme for protein synthesis in which DNA first serves as a template for nucleus-located synthesis of RNA. This RNA then in turn moves to the cytoplasm where it functions as the template for protein synthesis.

No one then had any compelling reason to take my hypothesis seriously, but by November 1952 I liked it well enough to print DNA→RNA→protein on a small piece of paper that I taped on the wall above my writing table in my rooms at Clare College. From the day of our first meeting, Francis Crick and I thought it highly likely that the genetic information of DNA is conveyed by the sequence of its four bases. But we knew it was premature to promote this idea before the structure of DNA was known. However, the moment we first saw how to build a double helix out of the four base pairs, it was clear that the essential uniqueness of a gene must reside in its respective sequence of base pairs. Moreover, not only could base pairing provide the way for genes to be exactly copied during gene duplication, but also could very likely underlie the process by which the genetic information of a DNA molecule is transferred to its RNA product.[1]

Still totally unclear, however, was how RNA might serve as the template for ordering the amino acids in their respective polypeptide products. Emboldened by our fantastic good luck in so simply finding the structural essence of the gene duplication process, I saw no reason not to

[1]Watson is discussing here messenger RNA (mRNA), which is copied from one strand of the DNA. Because there is no complementary strand it has nothing with which to pair off to form a double helix; but the molecule is flexible enough that runs of nucleotides can bend round to find any regions of complementary sequence, present by chance elsewhere in the chain, or form hairpin-like segments of double helix. The second major type of RNA present in the cell is ribosomal RNA: this, in association with a cluster of proteins, makes up the ribosome—the "reading heads," which scan the message along the mRNA and thereby direct the synthesis of the corresponding protein chain. The third type of RNA is transfer RNA (tRNA), the "adaptor" that carries the constituent amino acids to the site of protein synthesis (the ribosome).

take on the challenge of finding out what RNA molecules looked like in three dimensions. Such knowledge, I felt, would be indispensable to understanding how they functioned in protein synthesis. This task I took on when I moved on to Pasadena in the fall of 1953. There I joined forces with Alex Rich, who had started working on DNA just before we found the double helix. There was no difficulty in getting him to move on to a better pasture, and soon I was collecting RNA samples and drawing them into fibers that Alex exposed to X-ray beams. But despite much travail, even those fibers displaying high birefringence never gave rise to ordered diffraction patterns like those of DNA. Though we thought we saw reflections that might have come from short sections of double helices, we could never be sure and saw no way to decide whether RNA was a one- or two-chain molecule. Here the base composition was a bad tease. Viral RNAs clearly did not show equivalence of A with U or G with C, but the RNA from cells had A/T and G/C ratios sometimes closely approaching 1/1. But we could see no difference in the general features of the X-ray diagrams from viral or cellular RNAs. To our annoyance, RNA, no matter from what source, showed the identical X-ray diagrams characterized by strong reflections at 3.36 Å and 4.00 Å. Clearly there was some ordered structure in RNA, but we saw no way to get to it. After some six months of such frustration, we gave up.

A very different approach to protein synthesis came from the very clever, Russian-born, theoretical physicist George Gamow, who was struck by the fact that the 3.6-Å distance between adjacent amino acids in extended polypeptide chains was very similar to the 3.4-Å separation between adjacent base pairs in the B form of the double helix.[2] Not then cognizant of RNA's primary role in protein synthesis, Gamow proposed that amino acids are directly ordered by contacts with the DNA base pairs, with the polypeptide products containing the same number of amino acids as there are base pairs in their DNA templates. To deal with the fact that the DNA has only four letters in its alphabet while there are twenty different letters (amino acids) used to specify proteins, Gamow assumed that each amino acid must be coded by several adjacent base pairs. Because there are only 16 (4 × 4) combinations of the four bases, taken two at a time, Gamow made the assumption that each amino acid must be specified by groups of

[2]This structural correspondence turned out to be a mirage.

three adjacent base pairs (a codon) along DNA chains. To deal with the fact that there exist 64 (4 × 4 × 4) such triplets, he assumed that many amino acids must be specified by more than one triplet (redundant triplets). In such an "overlapping" code, adjacent amino acid codons have two out of the three base pairs in common, thereby restricting which amino acids can lie adjacent to each other. Gamow's first overlapping code was only one of several that later were devised, each leading to different combinations of forbidden amino acid neighbors.

When George first told me his scheme, I quickly dismissed it because DNA was not the template that ordered the amino acids. But he was having combinatorial fun and besides I could not rule out the possibility that some RNA molecules were double helices. With time, moreover, I realized there was a real virtue to Gamow-like codes. In disproving them, the possibility of overlapping codes could be ruled out, pointing the way to codes in which adjacent groups of most likely three bases specified successive amino acids along a polypeptide chain. In fact, the first known amino acid sequence, that of Sanger for insulin, disproved the first code although Gamow did not at first realize this, with his initial list of the 20 amino acids having several embarrassing mistakes (e.g., he included both cystene and cysteine). Other such overlapping codes devised over the next year were also ruled out as more proteins became sequenced. It was during the first coding rush that Leslie Orgel and I, on a trip to Berkeley, where Gamow was spending the spring of 1954, suggested that we form a club of twenty members whose purpose was to crack the RNA structure and in so doing reveal how the genetic code operated. Soon to be known as the RNA Tie Club, with its members reflecting Gamow's eclectic taste, it never had a formal meeting. Nor did all its members ever cough up the money to purchase their RNA ties and tie pins bearing their respective amino acid code letters. Gamow's tie pin sported ALA (alanine) and mine PRO (proline). Much more important in the long run was the opportunity it provided to exchange ideas about the code through "notes to the RNA Tie Club."

Several of these communications, the most important of which came from Francis Crick, Leslie Orgel, and Sydney Brenner, later became incorporated into published manuscripts. Among these was the definitive disproof by Sydney Brenner of any form of overlapping code, written in the fall of 1956 in Johannesburg, just before he returned to England to join Francis Crick. Equally important was the communication by Crick, John

Griffiths, and Leslie Orgel of a commaless code as a device to let nonoverlapping triplets be read in the appropriate reading frame. But the intellectually most influential paper was the RNA Tie Club's first note, written by Francis Crick and sent out early in 1955 under the title *On Degenerate Templates and the Adaptor Hypothesis*. After spending the previous August in Woods Hole batting about potential codes with Gamow and Brenner, Crick began questioning the basic assumption that a nucleic acid template provided specific cavities complementary in shape and charge to the amino acid side groups. Here he argued that the specific parts of nucleic acid bases want to hydrogen bond and were not all suitable for forming cavities that could attract the hydrophobic side groups of amino acids like valine, leucine, or isoleucine. Equally tricky to imagine was any structural basis for degenerative codes in which many amino acid side groups are specified by more than one set of triplets. Faced with what he considered insuperable obstacles, he made the radical proposal that prior to peptide bond formation, amino acids are enzymatically attached to small "adaptor" molecules that have surfaces specifically tailored to bond to nucleic acid triplets. Here Crick suggested they might be tiny polynucleotides that base pair to RNA templates.

My reaction to the adaptor hypothesis was initially very negative even though I had spent the fall of 1954 futilely trying to fold RNA chains into shapes bearing cavities appropriate for the amino acid side groups. The adaptor idea seemed much too complicated to me ever to have gotten started at the origin of life several billion years ago. Even Francis had his moments of doubt, in fact, concluding his now famous Tie Club Note with the phrase, "In the comparative isolation of Cambridge I must confess there are times when I have no stomach for decoding." In fact, soon after I returned to Cambridge for the year beginning in June 1955, Francis and Alex Rich became immersed in building new three-dimensional models for collagen to compete with one earlier proposed by Linus Pauling. This in spite of the fact that Francis and I had often referred to collagen as the most boring of macromolecules.

Equally out of frustration, I reverted to thinking about plant viruses—in particular, tobacco mosaic virus (TMV)—whose helical construction I had worked out in the spring of 1952 after Francis and I had been told by Sir Lawrence Bragg to stop trying to work out the structure of DNA through model building. Always troublesome to me was the apparent necessity to postulate both genetic and protein synthesis roles for RNA. Then, knowing that only a tiny fraction of TMV particles are actually infectious, I speculated whether in fact these rare particles contained DNA, not RNA, chains. But this possibility was ruled out when it became possible to reconstitute infectious TMV particles from their purified RNA and protein components. This experiment, first successfully accomplished in the spring of 1955 in Berkeley by Heinz Fraenkel-Conrad and Robley Williams, generated much newspaper publicity giving uninformed readers the belief that life itself had been created. Francis, however, put the matter in its proper light, being quoted in the English press that this was a finding he had anticipated.

Reconstitution by itself, however, did answer the question of whether the protein component played any more than a protective coat role for the genetic information bearing RNA component. Less than a year later, however, Alfred Gierer, working in Gerhard Schramm's lab in Tubingen, clearly showed that the RNA alone was infectious. This primacy of nucleic acids as bearers of genetic information then lay at the heart of the way Francis and I thought about cells and viruses. But this was far from an acceptable paradigm for many of the attendees at the key, late March 1956 CIBA Foundation meeting on the structure of viruses. They were not at home with the concept that information flows unidirectionally from nucleic acids to proteins and never backwards. This was awkwardly shown when Andre Lwoff and I passed on to Robley Williams a proposed telegram message from Wendell Stanley reading "TMV protein infectious—be cautious!" To our amazement, Robley didn't question the result until we told him the hoax.

At this gathering of some 30 scientists, Francis presented our ideas on why the protective coats of viruses are made up of protein subunits. In our view it was a consequence of the fact that no viral nucleic acid had sufficient coding capacity to specify a single polypeptide chain large enough to surround a more centrally located core of nucleic acid. The two million MW RNA of TMV, for example, contains only 6000 bases and so assuming a coding ratio of three bases per amino acid is only capable of specifying a 2000 amino acid polypeptide or about 230,000 MW. To make up the 38 million MW protein coat, at least 160 subunits would be needed. In fact, the TMV subunit contains only some 150 amino acids, suggesting that the TMV RNA codes for several proteins or that ratio is very much larger than three. Initially we thought that tomato bushy stunt virus (TBSV) with a much higher RNA content might be more useful in giving a realistic value for the coding ratio. It contains only four bases for every amino acid in the crystallographic subunit. But later direct analysis of its protein subunit size suggested that each crystallographic repeat contains some five protein subunits, very likely implying that TBSV RNA like TMV RNA also codes for several proteins.

After that meeting I again had a go at the RNA structure, taking advantage of the newly discovered enzyme polynucleotide phosphorylase, which Marianne Grunberg-Manago and Severo Ochoa found could be used to make synthetic RNA molecules.[3] By then back at NIH, Alex Rich had shown the month before that random AU and AGCU copolymers gave similar X-ray diffraction patterns to those that we had obtained using purified cellular and viral RNAs. I focused instead on Poly A (adenine) fibers drawn from material prepared in the Molteno Institute by Roy Markham and David Lipkin. To my delight, they generated clean helical X-ray diagrams that were best interpreted as base-paired double helices built up from two parallel Poly A chains. Initially I was disturbed by the fact that many of the key reflections overlapped with those generated by purified RNA, which by then we had every reason to believe was single-chained. Later this apparent paradox was possibly resolved by the finding that sections of hydrogen-bonded hairpins form along most RNA chains. Conceivably it is these short sections of imperfect double helices that generate the DNA-like feature of the RNA X-ray patterns.

By then discouraged that study of purified RNA would lead toward understanding how protein synthesis occurs, I decided to concentrate on the structure of ribosomal particles, believing that they must carry the genetic information for ordering amino acids in proteins. That spring of 1956, I had convinced Alfred Tissieres, then a Fellow at Kings working in the Molteno Institute on oxidative phosphorylation, to join me at Harvard, to which I would be moving in the fall of 1956. Alfred, in fact, had done some preliminary experiments on the then-called microsomal particles and was keen to follow them up.

---

[3] It proved possible to synthetically prepare RNAs in which all the nucleotides are the same (all U, say, or A), or which have random sequences of any two, three, or all four nucleotides.

I arrived at Harvard some six months before Alfred, preoccupied most of this time trying to be an effective teacher for the seniors and beginning graduate students who were taking my course on viruses. As soon as my lectures were under control, I went across the Charles River to see Paul Zamecnik, whom I had first met the year before while briefly stopping at Harvard on my way back to England. There in the building where Fritz Lipmann[4] also had his lab, I first appreciated the importance of the discovery (made there a year earlier by Mahlon Hoagland) of the activated high-energy acyl-amino acid intermediates for protein synthesis. Even more important, I first learned of the more recent observation (by Mahlon, Marjorie Stephenson, and Paul) of a soluble RNA (sRNA) fraction to which the activated amino acids are transferred prior to protein synthesis. Quickly I realized that these sRNA molecules might be the polynucleotide adaptors postulated two years before by Francis in his first note to the Tie Club. Prior to my visit, Crick's ideas were unknown to the Massachusetts General group, and they eagerly brought out Francis when they came together at the 1957 Gordon Conference on Nucleic Acids and Proteins.

Tissieres and I commenced our molecular characterization of E. coli ribosomes in the spring of 1957, suspecting that they might have structural plans like those of the small RNA viruses, whose isocahedral-shaped protein shells are formed by the regular aggregation of a single protein building block. We could not have been more wrong about how they are organized. To start with, we found that the E. coli ribosomes, like those from all other organisms, are formed by the aggregation of two RNA-containing subunits, with the larger 50s subunit approximately twice the size of the smaller 30s subunit. Each subunit contains a single major RNA chain, with the 50s ribosomal subunit possessing 23s RNA chains and the 30s subunit 16s RNA chains. Moreover, both contain a large number of different small proteins that for the most part are subunit specific. At low Mg^{2+} levels, the 30s and 50s subunits do not associate with each other, but when the Mg^{2+} concentration is raised they come together to form the 70s complex that subsequent work has shown to be the ribosomal form that carries out protein synthesis.

---

[4]Fritz Lipmann, German born and trained, was a Nobel Laureate and one of the most eminent biochemists of the twentieth century.

Naively, at first, we assumed that either the 16s RNA or the 23s RNA or both were the actual templates for protein synthesis. Puzzling, however, to us was why the templates existed in only two size classes while there was great variation in the sizes of their putative polypeptide products. Equally disturbing was why the base composition of ribosomal RNAs barely varied between bacterial species with highly different AT/GC ratios. A priori we had expected to find that the base composition of the RNA templates would reflect that of their DNA templates. Luckily there was one powerful exception. The phage T2-specific RNA made after T2 infection of *E. coli* was shown by 1956 by Volkan and Astrachan to have a T2 DNA-like base composition. Moreover, in contrast to the metabolically very stable ribosomal RNA chains, T2 RNA had been found to have a half-life of only several minutes.

In retrospect, Tissieres and I should have gravitated early to T2 RNA, but in fact not until the fall of 1959 were its molecular properties anywhere investigated. Then Masayasu Nomura and Ben Hall, working with Sol Spiegelman at the University of Illinois, provided tentative evidence for the incorporation of T2 RNA into abnormally small ribosomes. In trying to follow up this observation early in 1960, my graduate student, Bob Risebrough, came to a radically different conclusion. After T2 RNA is synthesized, it does not become part of a ribosomal subunit by aggregating with newly made ribosomal proteins. What in fact happens is that in the presence of Mg^{2+}, the T2 RNA becomes attached to the smaller 30s ribosome subunit, which in turn binds the larger 50s subunit to form the >70s complex that actually carries out protein synthesis. This result instantly changed the way we visualized protein synthesis. Instead of serving template roles, ribosomes function as stable assemblage sites for protein synthesis. The true template had to be a new RNA class unknown until that moment both because it comprises such a small percentage of the total RNA and because it is heterogeneous in length. Later these metabolically unstable templates, whose amounts respond to cellular needs, would be named messenger RNA (mRNA) by Jacques Monod and Francois Jacob.

The first lab outsider to whom I revealed this conceptual breakthrough was Leo Szilard, whom I had gone down to see in New York in early April 1960 where he was successfully plotting the radiation therapy that would cure his bladder cancer. Leo's reaction was entirely negative, not being convinced that we had the right interpretation for Risebrough's experimental results. Predictably he wanted us to show that messenger RNA existed in uninfected *E. coli* cells before he would change his mind set. These were experiments, in fact, that we had already planned to start several weeks later, as soon as Francois Gros would arrive from Paris to spend several months working with Alfred and me. Also soon to join this effort was Wally Gilbert, then still teaching theoretical physics to Harvard students, but who increasingly was tempted by our excitement about mRNA to move into molecular biology. By the time of the June Gordon Conference on Nucleic Acids, we were virtually convinced that *E. coli* mRNA also existed, and by the summer's end we had data for a convincing publication. Already at the Gordon Conference we heard rumors that Sydney Brenner and Francois Jacob, using very different arguments, had also postulated the existence of mRNA, and their idea was being tested by Sydney that week in Matt Meselson's lab at Caltech. Eventually, we were to publish our independent proofs for mRNA's existence early in 1961 in two back-to-back articles in *Nature*.

With the basic scheme for how RNA participates in protein synthesis known, the path became open for definitive experiments on the exact nature of the genetic code. By using enzymatically synthesized RNA as messengers for in vitro protein synthesis, by early 1966 all the triplet codons had their correct assignments. With this major goal achieved, the time had clearly come to ask how the DNA→RNA→protein flow of information ever got started. Here Francis was again far ahead of his time. In 1968 he argued that RNA must have been the first genetic molecule, further suggesting that RNA, besides acting as a template, also might act as an enzyme and in so doing catalyze its own self-replication. How right he was!

Bragg's Foreword to The Double Helix

1990

I had come to England for the fall of 1965 to finish *The Double Helix*. Harvard had granted me a sabbatical leave and the Guggenheim Foundation one of their prized fellowships. Sydney Brenner had arranged for me to live in King's in rooms looking out to Clare, and I had virtually completed the manuscript by mid-December. At that time a one-day meeting was held at the British Biophysical Society at Queen Elizabeth College located off Campden Hill Road in Kensington. While listening to the talks I was diverted by a very good-looking young woman to whom during the first coffee break I introduced myself. She was Louise Johnson, a crystallographer at the Royal Institution working for her Ph.D. with David Phillips on the structure of lysozyme. We went to lunch together at a nearby pub and without asking I got the unmistakable impression that she was fond of a man that she did not wish to identify. Her beauty, however, was not easily dismissable, and when I was next in London I arranged to stop by the R.I.

Then in mid-January I was on my way to Kenya for six weeks in East Africa, lecturing to students under the auspices of the Ford Foundation. The last chapter of *The Double Helix* had been finished while I was back in the United States in early January. Tom Wilson, the director of Harvard University Press, was among the first to read it and immediately said he wanted to publish it, provided that his lawyers could reassure him that no one was libeled, even better, if he could also obtain permission from key

33

figures like Francis Crick, Maurice Wilkins, and Sir Lawrence Bragg.[1] I knew that some awkward moments lay ahead of me.

Immediately upon arriving at the Mayfair Hilton, I phoned Peter Pauling, then a Lecturer in Chemistry at University College, asking him who was the beauty at the R.I., the site where he had done his Ph.D. I had no doubt he would know, and we arranged to meet the next day. Just before noon we called in at the R.I. to see if we could get Louise and others of her group that Peter knew well to join us for lunch. Happily she accepted as did Tony North, whom I first knew at the Cavendish. Soon we were walking down Dover Street and into Wheelers, where I explained the awkward moments that I must face soon when I gave my manuscript to Sir Lawrence.

Possibly helping me was the fact that Sir Lawrence had written to me at Harvard almost a year before, suggesting that before my recollections fade I should write up my detailed memories of the finding of the double helix. On the other hand, he was bound to find himself portrayed in an unfavorable manner in the early chapters. After we were on to our second bottle of chablis, Tony North came up with the initially seemingly perverse idea of my asking Sir Lawrence to do the Foreword. Quickly, however, everyone saw that this scheme would be the perfect way to give Bragg the respect he deserved without destroying my attempt to tell the story as if it were a novel, as opposed to a more conventional autobiography.

Upon my return from Kenya I was to be at Alfred Tissiere's lab in Geneva for several months. Soon I flew from there to London to give Sir Lawrence a copy of the manuscript. After arranging an appointment with his secretary, I went up to his flat within the R.I. Graciously he told me that

---

[1]Sir Lawrence Bragg, Cavendish Professor of Physics at Cambridge, had shared the Nobel Prize with his father for the discovery of X-ray crystallography. This became the primary, for many years the only, method of working out the structure in space of molecules. Its apotheosis came with the determination of structures of giant molecules, the proteins. Bragg (who received news of his Nobel Prize while serving in the trenches in France in 1915) was director of the Medical Research Council Unit, in which Watson and Crick made their discovery. On his retirement from the chair in Cambridge he became head of the Royal Institution, the laboratory in London, where Watson called on him. (Bragg was a genial character, who, frustrated that the Director's flat at the Royal Institution off Piccadilly lacked a garden, took to plying his spade as a part-time jobbing gardener. He was found out when a visitor to a house where he was working asked the startled owner what Sir Lawrence was doing in her garden.)

he wanted me to tell my side of the story because, given Francis Crick's brilliance, my contributions might well be thought those of a minor contributor. I told him that what he was to read would not be at all what he had asked for. My aim was to write an account where the characters as first portrayed were not always what they later turned out to be. So I was concerned he might not like the way I first introduced him. If, however, I were to describe his interactions with Francis in any way other than Francis described them to me, my book as a work of literature would be badly compromised. Our meeting lasted less than a half hour, it being arranged that I would return the following week upon my return from a visit back to Cambridge.

I was naturally nervous when I approached his office the second time. But I began to relax when he greeted me with the statement that he could not sue me for libel if he wrote the Foreword. Telling me that he was at first very upset by what he read, and had so told his wife Alice, he had calmed down and saw what I was trying to do. Clearly he appreciated the virtue of his writing the Foreword. He would be acting in a magnanimous as opposed to petty way and show that he was above flattery. With his consent in hand I flew to Geneva, knowing that a great hurdle had been crossed, though I had to fear that Sir Lawrence might change his mind if he found that others featured in the book thought its publication should be blocked.

Tom Wilson was equally relieved when later told of Bragg's reaction. Immediately he wrote to Sir Lawrence that Harvard University Press did indeed want to publish my book, which still at that time I called "Honest Jim" knowing of the past literary successes of Lord Jim and Lucky Jim. Bragg's Introduction arrived at Harvard in the mid-fall of 1966. By then we were clearly apprehensive that a serious attempt would be made to block its appearance. Francis Crick did not think it was academic enough for a university press, and President Pusey of Harvard, fearing the scandal of being caught up in a fight between noted scientists, told the Press not to publish it. Happily, Tom Wilson still ended up with the book because, knowing of his imminent retirement, he had arranged to move on as a senior editor to Athenaeum, a publisher of serious literature in New York. I worried that Sir Lawrence might back off during the ensuing controversy, but he held his ground and the book was published as *The Double Helix* in New York in February 1968. The English publisher was Weidenfeld and

Nicolson, whose edition appeared in May 1968. Six weeks earlier I was married to Elizabeth Lewis and we arranged later to visit Sir Lawrence and Lady Bragg at their home near the Suffolk coast. This was to be the last occasion we ever met, and I was most touched by his pleasure in showing us his garden that so pleased him, as did those of his previous homes on West and Maddingly Roads in Cambridge.

In retrospect I do not know whether I would have had the courage to see the publication of *The Double Helix* through to its end without Sir Lawrence's backing. In writing it, I thought science would be helped by its appearance. But certain of my friends, and most definitely my father, worried that I might be in for more trouble than I could handle. It meant much to me in the prepublication years to have behind me a man of such integrity and intelligence.

Biographies:
Luria, Hershey, and Pauling

Salvador E. Luria (1912–1991)

S.E. Luria, who died on 6 February, brought bacteria and their viruses, the bacteriophages, into the forefront of research on the nature of the gene.

Born in Turin, Italy, Luria initially trained to be a physician and chose radiology for his speciality. After military service he moved to Rome to learn more physics in the circle of Enrico Fermi and Edoardo Amaldi. There Franco Rasetti excited him about radiation biology and the formulations of the gene as a molecule by the German physicist Max Delbrück. Equally importantly, he discovered the existence of bacteriophages and began experiments on them in the laboratory of the bacteriologist Geo Rio. Growing official antisemitism in Italy necessitated his move to Paris in late 1938, where he continued working on phage under the patronage of the physicist Fernand Holweck, doing experiments that aimed to determine the sizes of phages through measurements of the rate at which they were inactivated by ionizing radiations.

The German conquest of France forced him again to flee, this time to New York City, which he reached in September 1940 by boat from Lisbon. At the suggestion of Fermi, Luria approached the Rockefeller Foundation, and the fellowship they provided allowed him to resume phage work at the College of Physicians and Surgeons of Columbia University. Soon he was in contact with Delbrück, also by then a refugee from totalitarian tyranny. By the summer of 1941, they had begun joint experiments on *Escherichia coli* and its phages, first at Cold Spring Harbor and later at Vanderbilt, to which Delbrück had moved to teach physics.

Initially, they focused on the interference phenomenon, which permits only one type of phage to multiply in a given bacteria. In the course of their experiments, they used bacteria variants resistant to specific phages. At that time it was an open question whether these resistant cells arose through gene mutations, because traditional wisdom among bacteriologists was that bacteria lacked chromosomes and a form of heredity similar to that of higher organisms. In particular the English physical chemist Cyril Hinshelwood argued against mutational origins of resistant bacteria, believing that they arose through altered chemical equilibria.

It was Luria who provided the key idea to distinguish between the mendelian and lamarckian explanations. What was needed was a way to show that the phage-resistant variants of *E. coli* existed before they came into contact with phage. In February 1943, soon after his move to Indiana University, he realized that the distribution of resistant bacteria in a series of different cultures should settle the question. If the bacteria were made resistant by contact with phage, the number of resistant bacteria would be very similar in all the cultures. But if they arose by mutations, they should be clustered in families, the size of which reflected the time at which the mutational events occurred during the growth of the cultures. Depending upon when the mutations took place, there could be one, two, four, eight, and so on, resistant cells in each culture tube.

As soon as Luria found his first evidence favoring gene mutations, he wrote to Delbrück, who then worked out the appropriate mathematical equations. The resulting Luria–Delbrück manuscript, which appeared late in 1943, changed the face of genetics by making bacteria the obvious organisms for research on the nature of the gene. Experiments could be done in a day instead of weeks, and billions of cells could be examined in searching for mutants and determining mutation rates.

The following year Luria went on to show that phages also spontaneously mutate as they multiply, giving rise to variants every bit as stable as those found in bacteria. Two years later, Delbrück and Alfred Hershey independently provided evidence for genetic recombination in phage, while Joshua Lederberg and Edward Tatum demonstrated genetic recombination in *E. coli*. Thus, within a brief three-year span, the existence of chromosomes within both *E. coli* and its phages was established. For these experiments and Hershey's subsequent demonstration that DNA was the genetic component of phages, Luria, Delbrück, and Hershey in 1969 received the Nobel Prize in Medicine and Physiology.

Less appreciated was the key role played by Luria in the 1952 discovery of the host-modification phenomenon in which the host range of a progeny phage can be influenced by the exact strain of bacterium in which it has multiplied. A decade later, Werner Arber went on to show that such behavior reflects enzymatic attack on unmodified phage DNA and its prevention by methylation.

Luria was an exceptionally talented writer. His scientific papers, textbooks, and books for the general public all reflect mastery of his adopted English language. His first popular book, *Life: The Unfinished Experiment* (1973), won the National Book Award. His autobiography, *A Slot Machine, A Broken Test Tube* (1984), lucidly describes his intellectual and humanistic development. I remember him to be a teacher of the first rank. In autumn 1947, after only a few days into his course on viruses, I wanted to do my Ph.D. under his supervision. It was typical of his devotion to his students' future success that he later arranged with John Kendrew for me to go to the Cavendish Laboratory where I was to meet Francis Crick.

Equally important was his skill as an administrator, shown over the thirteen years he wisely and compassionately directed MIT's newly formed Cancer Center.

He was a human of passionate political beliefs. From early youth he identified himself with the causes of organized workers as opposed to those of management. As the Vietnam War developed, he became increasingly prominent in the antiwar movement and refused payment of the war-related component of his income taxes. Oppression of any form disturbed him, and there was never any doubt as to where he stood about authoritarian governments or institutions. More recently he worried about society's handling of the human genetic data whose accretion would be greatly accelerated by the human genome project. The possibility of a genetically defined underclass disturbed him, and we disagreed as to whether our societies would find the means to protect the victims of unjust throws of the genetic dice.

Luria knew what science at its best was and strove at all times to maintain high standards in both science and the human behavior that make it possible. In doing so while young he could offend out-of-date minds or those who were all too self-satisfied with their accomplishments. By the time he died, however, there were few who did not feel better by being in his presence.

Alfred Day Hershey (1908–1997)

Much of the basic facts about the gene and how it functions were learned through studies of bacteriophages, the viruses of bacteria. Phages came into biological prominence through experiments done in wartime United States by the physicist Max Delbrück and the biologist Salvador Luria. They believed that in studying how a single phage particle multiplies within a host bacterium to form many identical progeny phages, they were in effect studying naked genes in action. Soon they recruited the American chemist-turned-biologist Alfred Hershey to their way of thinking, and in 1943 the "Phage Group" was born. Of this famous trio, who were to receive the Nobel Prize in 1969, Hershey was initially the least celebrated.

Al had no trace of Delbrück's almost evangelical charisma or Luria's candid assertiveness and never welcomed the need to travel and expose his ideas to a wide audience. He framed his experiments to convince himself, not others, that he was on the right track. Then he could enjoy what he called "Hershey Heaven," doing experiments that he understood would give the same answer upon repetition. Although both he and Luria had independently demonstrated that phages upon multiplying give rise to stable variants (mutants), it was Hershey, then in St. Louis, who in 1948 showed that their genetic determinants (genes) were linearly linked to each other like the genes along chromosomes of higher organisms.

His most famous experiment, however, occurred soon after he moved to the Department of Genetics of the Carnegie Institution of Washington in Cold Spring Harbor on Long Island. There, in 1952, with his assistant, Martha Chase, he showed that phage DNA, not its protein component, contains the phage genes. After a tadpole-shaped phage particle attaches to a bacterium, its DNA enters through a tiny hole while its protein coat remains outside. Key to Hershey's success was showing that viral infection is not affected by violent agitation in a kitchen blender, which removes the empty viral protein shells from the bacterial surface.

Although there was already good evidence from the 1944 announcement by Avery, MacLeod, and McCarty at the Rockefeller Institute that DNA could genetically alter the surface properties of bacteria, its broader significance was unknown. The Hershey–Chase experiment had a much stronger impact than most confirmatory announcements and made me ever more certain that finding the three-dimensional structure of DNA was biology's next most important objective. The finding of the double helix by Francis Crick and me came only 11 months after my receipt of a long Hershey letter describing his blender experiment results. Soon afterward, I brought it to Oxford to excitedly read aloud before a large April meeting on viral multiplication.

Hershey's extraordinary experimental acumen was last demonstrated through his 1965 finding that the DNA molecule (chromosome) of bacteriophage λ had 20-base-long single-stranded tails at each end. The base sequences of these tails were complementary, allowing them to find each other and circular λ DNA molecules. This was a bombshell result because circular molecules had two years before been hypothesized as an intermediate in the integration of λ DNA molecules into bacterial chromosomes.

Key to the *esprit* of the phage group was its annual late summer meeting at Cold Spring Harbor. It was not for amateurs, and the intellects then exposed had no equal in biology then or even now, in the hurly-burly rush of genetic manipulation of today, when perhaps a quarter-million individuals think about DNA in the course of their daily lives. Logic, never emotion or commercial consideration, set the tone in those days, and it was always with keen expectation that we awaited Al's often hour-long concluding remarks. Tightly constructed, his summaries struck those of us aware of his acute taciturnity as containing more words than he might have spoken to outsiders in the whole of the past year.

By 1970, the basic features of λ DNA replication and functioning, both in its more conventional lytic phase and in its prophage stage, had become known, and a book was needed for their presentation to a broader biological audience. Because of his stature and honest impartiality, everyone wanted Al to be the editor and to see that the fifty-two different papers said what they should and no more. Through Al's ruthless cutting of unneeded verbiage, the book's length was kept in check and the final volume was not painful to hold in one's hands. Working hard to make the changes that Al suggested, the young Harvard star, Mark Ptashne, noted with pleasure that Al had made no further changes in his revised manuscript's first ten pages. Then on page 11, Al wrote, "Start here." Only Al could be so direct and so admired.

This was his last scientific hurrah. Although he was only 63, he soon chose to retire. It bothered me that a mind so focused and inventive would willingly stop doing science, but he lived to his own standards, and the pursuit of new ideas was never easy. New moments of Hershey Heaven never lasted long, and his past long summers away from the lab sailing on Georgian Bay were out of place as the number of new scientists seeking gene secrets increased. In retrospect, he did not get out too soon. Recombinant DNA was but two years away, and soon there would have been competitors on all sides.

Retirement saw him first expanding his garden, and I could exchange words with him on my walks past his home. Later, when he became absorbed with computers, he was inside and I never thought I had something important enough for an excuse to interrupt. I now regret my lack of courage. Al always appreciated others trying to move ahead. In his last year, much curved over by arthritis, Al drove his wife to my house for a small gathering. In asking about our daily lives at the lab, Al likely knew this would be the last time we saw him. The time to go was at hand, and he stopped eating. No one among his friends ever expects to see another who so pushed science to that level of human endurance.

Linus Pauling (1901–1994)

In 1931, at only thirty years old, the Oregon-raised Linus Pauling knew he was the world's best chemist. Ten years later the rest of the world's chemists agreed. They marveled at his use of the new quantum mechanics of the European theoretical physicists. Already Pauling's masterful 1939 book, The Nature of the Chemical Bond, was on its way to being chemistry's most influential book of this century and its effective bible.[1]

Linus's charismatic self-confidence as a young professor, however, appealed more to those below him in age than to those who controlled the funds at the even younger California Institute of Technology. Luckily the

[1] *The Structure of the Chemical Bond*, the first edition of which appeared in 1939, is probably the most influential book ever published on chemistry. Its influence on biochemists was equally profound. A colleague of the great biochemist, Fritz Lipmann (see p. 30) once said that every time Lipmann read another chapter of Pauling's book he made another discovery that changed biochemistry.

Rockefeller Foundation, then the most powerful American source of money for science, came to his rescue. They allowed him to begin thinking about proteins. Then virtually all other chemists thought these macromolecules too complex to study. His biggest success in biology came from his 1951 proposal of the α-helical fold for proteins.[2] With it quickly verified in England, Linus's confidence was never higher.

Then unexpectedly he struck out. In late 1952 he proposed an implausible three-chain helix for DNA. Soon after, in Cambridge, England, Francis Crick and I, apprehensive that Pauling might quickly bat again, found the double helix. Why Linus failed to hit this home run will never be known. Pressed too often by outsiders, his wife, Ava Helen, told Linus that he should have worked harder. Perhaps more important, the decade that followed the war's end had too many moments of personal agony.

They arose from Pauling's opposition to nuclear weapons. After the first atomic bombs were used, he began giving speeches, feeling that our nation's growing anti-Communist fears were forcing us to an insane nuclear weapons race. Broadly labeled a "pink," if not a "red," J. Edgar Hoover personally pursued him, Senator McCarthy called him a security risk, the State Department took away his passport, and many of its trustees wanted Caltech to fire him.

Concerned that his university was damaged by his visibility, Linus resigned from several blacklisted organizations and stopped giving political speeches. But he still could not get a passport, and our government would not give him research grants. Gradually he came back to the political lecture circuit, saying that science cannot flourish when scientists are punished for saying what they believe. Emboldened by his 1954 Nobel Prize in Chemistry, he emphasized dangers from the radioactive fallout

---

[2]The discovery of the α-helix, a fundamental element in the structure of proteins, was the first great triumph of model-building. Pauling told of how he did it: While on a sabbatical visit to Oxford, he was confined to bed with a heavy cold. Bored with his detective novel, he amused himself by cutting out the form of a protein (polypeptide) chain from a sheet of paper and he then twisted it to make a fit between the units along the chain. His success caused chagrin in Cambridge, where Bragg (p. 34) and his colleagues had tried and failed to produce a convincing model (because they had handicapped themselves by imposing the invalid rule that there must be an integral number of units in each turn of the helix). Competition from Pauling was what Watson and Crick most feared when they were attempting to construct a model of DNA, as Watson so entertainingly relates in *The Double Helix*.

emanating from the new superbombs being exploded in the Pacific and in desolate regions of Russia.

Only needing $250 for postage, he organized petitions from scientists throughout the world to ban nuclear tests. Worried by his effectiveness, *Time* magazine ran his photo over the caption, "Defender of the unborn or dupe of enemies of liberty." When three prominent Caltech trustees resigned, President DuBridge showed that their voices still mattered by accepting Pauling's resignation as Chemistry Department Chairman and cutting his salary from $18,000 to $15,000 per year. Never publicly complaining, Pauling later took comfort in President Eisenhower's beginning negotiations for a nuclear test ban. Finally, under President Kennedy, a comprehensive atmospheric test ban treaty between Russia and the United States was signed in the summer of 1963.

Though Pauling's scientific arguments about fallout risks were highly speculative, equally shaky were the claims of the physicist Edward Teller that "clean hydrogen bombs" could be developed by further tests. Both he and Linus were playing politics toward what they thought were nobler ends. In retrospect, Linus saw the bigger picture when he was appalled by the large number of nuclear weapons that we and the Russians were creating.

For his test ban activities, the Nobel Peace Prize was awarded to Linus in late 1963, making him the only individual ever to receive two unshared Nobel Prizes. No true congratulations came from the Caltech hierarchy and, at the age of 63, Pauling resigned from its faculty, 41 years after he came to Pasadena as a graduate student. There would have been no Caltech party to say good-bye except for a last-minute one held by its Biology Department.

Linus's last big wish was to do for medicine what he had done for chemistry and biology. But using vitamins to conquer mental diseases, the common cold, and cancer proved more than a tall order even for Pauling. Today these terrible afflictions remain much unchecked. Near his death from cancer at 93, Linus said that he would have died sooner were it not for his massive daily doses of vitamin C. I prefer the conclusion that vitamin C does not kill.

That Linus did not get his final triumph should not surprise us. Failure hovers uncomfortably close to greatness. What matters now were his perfections, not his past imperfections. As President Clinton wrote to

Pauling's children upon learning of his death, your father's "tireless activism for the cause of peace helped force international leaders to reassess their priorities, ultimately making our world a safer place."

Now I most remember Pauling from fifty years ago when he proclaimed that no vital forces, only chemical bonds, underlie life.

Recombinant DNA Controversies

In Further Defense of DNA

1978

I feel most honored to be giving the Lynen Lecture,[1] because I am the first lecturer not trained as a biochemist—I take this to mean that my students have practiced it well. When Professor Whelan finally cornered me to say whether I would accept the invitation, I had to admit I was deeply tempted because it was to be autobiographical, and that might mean a little gossip. But then he said I had to write it up, and I feared the lawyers would again be in my life. But finally I thought, "Well, I might as well give it, because if I don't they might ask Bob Sinsheimer who is afraid of recombinant DNA." So now I wish to relate why I got in my present position running a lab with an interest in tumor viruses, and how in the process I became a minor actor in our current drama about DNA.

The best time to start is in the late winter of 1958 when Salva Luria invited me to give the Miller lectures at the University of Illinois. It was cold, and living at the Union was not exactly fun. But I met Nomura, who then was a postdoc with Spiegelman, and that led to Masayasu's spending the subsequent summer at Harvard. Sol's notebooks were unbelievably neat, and classical music was played at all times. One evening Van Potter showed up and spoke on cancer before Gunsalus and his biochemists. He was thinking along the lines of the new ideas of Umbarger and Pardee on feedback inhibition, a concept to which I had not previously paid much attention. I got quite excited and returned to Harvard thinking I should someday give a course on cancer.

[1]The lecture is named to commemorate Feodor Lynen, a leading German biochemist.

A year later Francis Crick arrived with Odile to be a visiting professor in the Chemistry Department for the spring term. He did the DNA→RNA→protein story for graduate students, while I gave it at the introductory level in Biology II. This was a new course and was to provide beginning Harvard students with their first opportunity to learn the revolution in molecular genetics. I used my last lecture to talk about cancer, and soon afterward I became manic when I realized that Seymour Cohen's recent work might be the key to how viruses make certain cells cancerous. He had just reported that the T-even phages[2] had genes that code for enzymes involved in DNA metabolism. If this was also true for animal viruses, the proliferative response of cells to viral infection might fall out. Unlike bacteria, most cells in a higher animal are turned off for DNA synthesis. So many, if not all, animal viruses might need to carry genes that would ensure the turning on of the cellular apparatus for DNA synthesis.

The following fall term I devoted my formal teaching to a course on the Biochemistry of Cancer, and for the first time I tried to master the literature. The Warburg effect was hard going, and I went down to NCI to talk with Dean Burk, then the most prominent scientist arguing on Warburg's side. To my surprise he had only a modest-sized lab and had virtually stopped active experimentation on why most tumor cells consumed so much glucose.

Soon after, in 1960, my student, Bob Risebourgh, was doing the first clean experiments showing that T2 RNA never became an integral part of any ribosomal subunit but was a separate species that bound to 70s ribosomes in high Mg^{2+}. It, not the ribosomal RNA, must be the template that orders amino acids during protein synthesis, and soon it became known as messenger RNA (mRNA).

The thought processes of the molecular biologists and our biochemist colleagues had for now become almost indistinguishable, and there was no longer any need to argue our respective cases. Only Chargaff wanted to fight. A unique opportunity came at the 1961 Cold Spring Harbor Symposium, the first time there was massive talk about messenger RNA (mRNA). Toward the end of a session, Sydney stood up to say that the term "messenger" was most appropriate for the new RNA form since it was a

---

[2]The T-even phages are a series of DNA-containing bacteriophages, named T2, T4, and so on.

very mercurial substance and Mercury was the "God of Messengers." Chargaff arose and asked if Dr. Brenner knew that Mercury was also the "God of Thieves."

Over the next five years, I increasingly pushed much of my lab onto Norton Zinder's newly discovered RNA phages, reasoning we should always study the simplest systems. At the same time I was more and more aware that Arthur Kornberg's lab consistently did better biochemistry than the rest of us. And everyone knew what you would do if you were in Arthur's lab—get in by 9 a.m. and purify your enzyme. So I encouraged my student, John Richardson, to purify RNA polymerase[3] better than anyone else, hoping that he would find evidence for specific binding to DNA. Though we learned much news about the physical properties of RNA polymerase, in retrospect all we observed were nonspecific interactions. That problem didn't break until 1969 when Andrew Travers arrived from Sydney's lab and he and Dick Burgess found σ. Most excitingly, σ disappeared from polymerase purified from T4-infected cells, and we thought that "At last we know the basis for the early vs late mRNA turn-ons; and even more important, maybe changes in RNA polymerase specificity will be important for embryology."

Ever since we had the double helix I had avoided thinking much about embryology. As a student I had a course on invertebrates with Paul Weiss, and despite being scared by his nasty frown, I enjoyed his *Principles of Development*. And later in Europe, I temporarily knew enough French to read eagerly the first edition of Brachet's *Embryologie Chimique*. But after the double helix emerged, I began to realize how difficult it would be to meaningfully describe at the molecular level even the most simple embryological process. No one with sense should anymore get excited by the observation that one substance goes up and another down during differentiation. We all knew this must happen. You could go from here to eternity measuring enzyme levels, but nothing would come out. Yet silly nonsense kept appearing from prominent mouths that soon we were going to have real embryological breakthroughs. So starting in 1959 I annually gave a lecture in Biology II against embryology and what was wrong with

[3]RNA polymerase is the complex enzyme responsible for copying the sequence of a DNA strand into the corresponding messenger RNA. One of the parts of the enzyme is the sigma factor, which is important in directing it to the right starting point in the DNA.

Woods Hole. I minced no words that if we were to get anywhere, we should stick to soluble problems. The Harvard biologists finally could not take me any longer and used my 1966 sabbatical to get me out of the course. From then on, Carroll Williams gave my lectures. I didn't really mind, because *The Molecular Biology of the Gene* had appeared, and I already had a much larger audience than I could warn against phony optimism.

It is not true now, but until all too recently most biologists, at least those I grew up with, were not that deep. They too often threw out corn instead of sense. As Americans they had to be optimistic, so why not be optimistic about embryology? But I thought it was a dead subject and should stay in the deep freeze. Common sense told me it would never really move unless there was a path back to DNA. But then there was no way to get out the DNA pieces you wanted to study. The eucaryotic genome was too big. If you wanted to be sensible and yet have future embryological ambitions, you worked with the DNA animal viruses.

Strongly influencing me was the recent discovery of the mouse virus polyoma. Its chromosome was almost as small as those of the RNA phages and yet it would make cells cancerous. Ever since Stoker and Dulbecco got the field going in the early 1960s, I wanted to go into it, at least in a vicarious fashion. When the Directorship of the Cold Spring Harbor Laboratory fell open, the opportunity arrived. John Cairns couldn't take the administrative chaos any more, and given the Lab's decrepit state, finding a new Director was not going to be simple. Its Board of Trustees did not know what to do, and Gunsalus wanted to offer the job to a phage geneticist located in Dallas. But I couldn't see it falling in the hands of someone with no emotional attachment to Cold Spring Harbor whom I feared would use it as a stepping stone back to Germany. Late in 1967 I said I would take it over on a part-time basis with the idea that I would build up a group that would do DNA tumor-virus molecular biology.

A few months later Liz and I were married, and as soon as the Harvard spring term ended, we came down to Long Island to spend the summer. The Lab had no free cash that anyone might consider spending, and much of July and August I spent throwing out junk from the library so that we might have a decent space to read journals. John Cairns told me to go after Joe Sambrook if I wanted to start animal virus work, and luckily Joe was to be about during part of Phil Marcus' and Gordon Sato's summer course on animal cells and viruses. By September I had decided to put together a grant request, and Lionel Crawford, who had come over for the summer to collaborate with Ray Gesteland, helped me draw up plans for converting James' lab into space for tumor virus work. By then Joe seemed eager to come here from the Salk, and the application went off to NIH and eventually ended up in the Genetics Study Section. One of its members was Charles Thomas, who knew that we necessarily would work with larger amounts of DNA than was handled by most nonmolecular tumor virologists. So he asked that a committee be set up to ask whether there might be a safety problem.

Until then there had essentially been no public discussion about possible biohazards of tumor virus work. Those scientists who grew up in the tradition of medical microbiology naturally knew that animal viruses should be treated as potentially dangerous. They could reassure themselves that animal virologists didn't have shorter life spans than the average, and Dulbecco and Stoker, for example, did their experiments on the open bench. Perhaps they had moments of mild apprehension when they first got into tumor virus work, but soon realized it made sense to focus their worries on well-documented hazards.

Committees never move fast, and we got the grant to start SV40 work without any restriction as to the possible biohazard. Our immediate aim was to get experiments going as fast as possible, and during our first year Joe and Bill Sugden focused on purifying eukaryotic RNA polymerases with the hope of turning up σ-like specificity factors. Like others before them, they found evidence for the I and II forms, but never had success with either in observing specific transcription. In the meanwhile, Carel Mulder was trying to find a restriction enzyme that would break SV40 at a specific site so that a linear denaturation map could be worked out by Hajo Delius, and maybe later the early and late mRNAs could be assigned to clean locations on the genome.

The potential biohazard problem surfaced again after Bob Pollack arrived here in 1971. Bob started out as a physicist and didn't like his former mentor, the trained M.D., Howard Green, who thought such talk was unwarranted. While medical students start out worrying whether they might get sick from a patient, you can't become a doctor unless you lose that fear and start practicing your trade. Bob, however, was a product of a Brooklyn socialist heritage which told you that worker's lives are often lost in the shuffle of their boss's success. He was eager to take on the biohazard dilemma and was a prime mover behind the first Asilomar gathering that took place in February 1973. It focused on the potential health hazards of tumor virus research, and data were presented, for example, on the SV40 contamination of the early Salk and Sabin vaccines. No general conclusion could emerge. There was not a trace of evidence that what we did was dangerous and would lead to any new cases of cancer. Yet maybe we hadn't waited long enough, because the incubation period of cancer can be very long. About the only solid impact was that NCI became likely to fund more elaborate containment facilities for work with tumor viruses and to sponsor courses on safer lab procedures. Mouth pipetting disappeared, and we tried to operate so that viruses would not go up in aerosols created by our experimental procedures. But it was impossible, and is still, to know whether our responses were adequate. No one had any idea of the magnitude of the possible dangers, and maybe there was no risk at all.

We could by now press ahead at increasing speed because the first useful restriction enzymes had been found and many more were soon to come from Richard Roberts' group in Demerec's lab. Early and late SV40 mRNAs were found both here and at NIH to be coded by different DNA strands, and each was found to occupy about one-half of the genome. The size assigned to the early region was just sufficient to code for the T (tumor) antigen, which Peter Tegtymeyer was estimating to be in the 80,000–100,000 range. An obvious next step was to purify this protein, because it was the only viral protein expressed in SV40-transformed cells and must be the oncogenic product that converts normal cells into their cancerous equivalents. Klaus and Mary Weber came down on leave from Harvard with this objective, hoping subsequently to find out how it works. Though Klaus is about as talented a protein chemist as you find anywhere, he found the problem more than he bargained for. After a year of hard work, he gave up because there was no way to get enough cells. T antigen is made normally in small quantities, and he was limited to work with cells that grow only in monolayers. Even if he had grown an order of magnitude more cells, the effort would likely have failed, and the possibility never opened up to work at the Kornberg level.

This problem was only to break several years later when Eugene Lukanidin came here from Moscow and, with John Hassel and Joe Sambrook, isolated new SV40-adenovirus hybrid in which the SV40 T antigen gene came under control of a late adenovirus promoter. This leads to a ten-fold-greater T antigen yield per cell. In addition, these hybrids, like wild-type adenoviruses, grow on cells that multiply in suspension to reach numbers far in excess of those limited to monolayers. Given these tricks, last year Bob Tjian used classical high-level biochemistry to show specific binding of T antigen to DNA fragments containing the origin of replication of SV40. I was very pleased, because our 1959 ideas on how DNA tumor viruses made cells cancerous were looking more and more correct.

Already by 1973 restriction enzymes were very key tools for deep probing of the viral genome, and they seemed likely to be even more so given the existence of Herb Boyer's and Stanley Cohen's new recombinant DNA technology. With plasmids to use as cloning vehicles, any piece of eucaryotic DNA eventually could be prepared in amounts suitable for detailed molecular characterization. We then found preparing enough viral DNA for chemical analysis a major task, and Wally Gilbert's and Fred Sanger's powerful new DNA sequencing procedures were yet to appear. So we immediately thought about using recombinant DNA technology to amplify tumor virus genes. In this way we wouldn't have to grow large amounts of virus anymore, and if biohazards did exist, they now could be greatly lessened.

But first Bob Pollack and then many participants at the 1973 Nucleic Acid Gordon Conference began to question whether we might inadvertently, or conceivably quite intentionally, use recombinant DNA technology to produce bacteria with extended pathogenicity. Questioning began as to whether in fact certain experiments should never be done. A close vote in favor of possible restraints was taken at the Gordon Conference, and the organizers led by Maxine Singer and Dieter Soll sent off a letter to *Science* magazine asking that the matter be taken up by a body like the National Academy. Phil Handler agreed and Paul Berg was asked to come in with a report. Subsequently he asked David Baltimore, Herman Lewis, Dan Nathans, Sherman Weissman, and me to meet with him when he came East to MIT early in 1974.

This was a period when the rights of innocent third parties were being taken increasingly serious. We should be more than fair to the dishwashers and technicians on whom we might blow viruses, or they on us, and so on. Motivated primarily by a desire to be maximally socially conscious and without any evidence that recombinant DNA was dangerous, we called for a partial moratorium until we had a big meeting the following

February. It became Asilomar II. In retrospect Asilomar I didn't produce anything except a small book, *Biohazards in Biological Research*, which Cold Spring Harbor put out at an almost give-away price and after several years made a tiny profit. I thought with Asilomar II we might have a book that would be a best seller. So to start with, I was quite enthusiastic about the whole affair. David Baltimore subsequently argued that people did not want the further hassle of still another meeting with manuscripts, and I got no book and stopped thinking about the matter. The thought never occurred to me that the new Asilomar could lead to any formal rules, because I found the subject too hypothetical for a reasonable response.

I could not have been more wrong. To start with, the National Academy organized a press conference at which analogies were made to the wartime decision of our nuclear physicists to desist from further publishing their experiments. This led to the press producing copy that we molecular biologists had a genie that we might not be able to contain, and that society might be at risk. None of us then thought it pertinent that we already take chances with DNA. Every time a new baby comes forth it is the product of new forms of DNA, and *a priori* we do not know whether the consequences will be for the good or bad. Recombinant DNA, *per se*, is not something first brought forth by science. It is an obligatory fact of life whose occurrence is far wider than generally perceived. Viruses, for example, have the potential to cross species barriers and to carry DNA between unrelated organisms. Then, however, it seemed more to the point to worry about what bad might come out of our labs rather than focus on the even more complex happenings that must occur naturally. That myopia, of which I also was guilty, will haunt us for a long time.

As soon as Asilomar II started, it became apparent that a hectic rush into guidelines was foreordained. Sydney Brenner set the tone by saying we must be responsible to society and face up to what our restriction enzymes might yield. I was not convinced and rose to say that because we were unable to bring forth any rational guidelines for tumor virus research, how could we now react to a situation where we could not even guess the size of our potential opponent? As far as I was concerned, everyone might as well go home. That response generated stony silence, and I was soon the meeting's outcast.

As the meeting progressed, its main preoccupation became the so-called safe strains. Waclaw Sylbalski first lit on this device, but it was Sydney who orchestrated its star role. Why not take K12 and make it so enfee-bled that it could never escape into the sewers of MIT? To most, this seemed to be almost the perfect response to the dilemma of not knowing whether we had something to worry about. But how much do you want to spend making a safe bug even safer (and to be sure K12 is safe)? Certain-ly twenty-five cents or even five dollars and maybe fifty thousand dollars. But would we do it if it were eventually to cost more than twenty-five mil-lion dollars and potentially lead to so many mucked-up memoranda of understanding that to cover our flanks we have to go to law school before risking recombinant DNA experimentation?

Here there is no point in recounting the three years' travail that have elapsed since Asilomar II. So many thousands of man-years have already been consumed that I would not further dwell on the matter were it now not so obvious that embryology acquired a future the moment we became able to clone specific pieces of eukaryotic DNA. The way is now open to find proteins that bind to specific pieces of DNA, and over the next decade we should come to understand key embryological steps the same way we now understand the lactose operon.

Today there is nothing to stop us except for the inability of biohazard experts to make rational responses to questions whose answers demand knowledge that we will never possess—or the agony of preparing count-less memoranda of understanding that will no longer be valid once you advise a clearer way to do your experiment—or the inability of the ever-growing NIH bureaucracy to know how to deal with novel experiments for which the guidelines have not yet been provided. All this is most mad-dening—especially because I now don't know of a single person who does recombinant DNA research who feels the tiniest apprehension. I don't think that a little Dictyostelium DNA in *E. coli* would ever make Harvey Lodish tremble even slightly. If it did, I would think him cuckoo.

Unfortunately, lots of well-intentioned outsiders today see recombi-nant DNA as a test case for the scientists' responsibility to society. Yet by now almost every molecular biologist wishes to expunge the guideline controversy from their consciousness. A return, however, to Asilomar II freedom will not be easy. I sensed that all too well as I made a short state-ment this past December before a public NIH hearing that Don Fredrick-son called to consider relaxing certain of the more onerous guidelines. My message to his advisory body was that it was a national disgrace that we were wasting our time with untestable speculations, and that the National Institutes of Health already had enough to do dealing with real human diseases. In response, a lawyer who represented the Ford Foundation-created environmental lobby, "The National Resources Defense Council," shrilly demanded that I prove that recombinant DNA research was safe. I looked across at him and said, "How do I know that you are not a paid killer sent by the Commies to do in our science?" He seemed stunned and did not reply. After the session ended, he came up to me and indignantly said, "How can you let me down? You scientists have created this issue, and you should keep it going." To which I replied, "Because I was a jackass is no reason for you to continue to be one."

Soon recombinant DNA is going to be back in front of Congress, and they will ask again whether there should be formal laws that will aim to reassure, say, the comfortable people of Princeton that its University's molecular biologists will not make them all sick. I focus upon the affluent because the more money you have, the more likely you have been frightened by DNA. Poor people do not know about DNA. You have to have leisure to be able to read and worry whether your community should permit DNA research. Only then will you learn that P4 level experiments are said to be potentially more dangerous than P3 and then need much persuading that P2 should go forward.

There will be no way, however, that we can ever make clear to our non-scientific associates why Asilomar II was so frightened by human DNA. Future work with it was declared possibly the most risky, and hence it is still virtually impossible to do anywhere. This was a strange decision and totally disregarded the fact that we have DNA in our sperm and eggs. Countless tons of human DNA daily already get spread about without our recombinant DNA technology, but this is not polite conversation. So laboratory rearrangements of human DNA are banned while normal and abnormal sex go on their uncontrollable ways.

Despite all this past confusion, the voices that the newspapers report still remain largely those who want to shut us down. To my dismay, the vast majority of our nation's biologists, the only group that has the competence to put its house back in order, stays largely mute. They seem not to remember that, historically, freedom is more easily lost than won and that the Director of NIH needs to hear from them as well as from the do-nothing fringe of the environmental movement. He must necessarily act as if he is not playing free and fancy with our nation's health. The only way he can do this is for the responsible scientific community to come to his aid and firmly say that an overblown issue should not stay overblown forever.

Five years have passed, and we have more than respected the right of all sides to be heard. Now in the absence of the slightest evidence that any danger to society is involved, we should go forward at the maximal possible speed. No longer are we a pastoral society, and we shall not survive well, if at all, without more science in our future. Instead, these days we too often play timid, if not mildly guilty, and then wonder why we are increasingly impotent in carrying along the public. As long as we whine and not come to the heart of the matter, we just look self-serving. To say that Asilomar II was science at its best, and that all would be fine if it had not been for Mayor Vellucci and his anti-recombinant DNA Cambridge City Council, is to miss the point. Our problem is not recombinant DNA, it is ourselves.

Standing Up for Recombinant DNA

1978

The collective decisions of knowledgeable men go sour more often than we want. We should not be surprised, however, that blue ribbon solutions like those proposed several years ago for recombinant DNA research frequently go astray. High-level committees are generally called into action only when a problem arises that has no simple response, or maybe no answer at all.

This was the case when in 1973 Herbert Boyer and Stanley Cohen came upon a method for rearranging DNA molecules in the test tube to create hybrid molecules ("recombinant DNA"), which might be derived, say, partly from mouse DNA and partly from bacterial DNA. Their ability to put these recombined molecules functionally back into cells had possibly opened the way to the creation of novel life forms that might alter the course of evolution. Moreover, the procedures for making recombinant DNA were simple and cheap, if uncontrolled, and were likely to be universally taken up and exploited throughout the world within a year or two at most. So it was natural that concern should develop as to whether we should just plunge ahead and hope for the best or whether we should try to block the momentum of recombinant DNA research until we could be sure that we were not doing ourselves in.

But whom to turn to was not at all obvious, because recombinant DNA research was so open-ended that no one person was clearly qualified to point the way. This unsettling fact should have warned us that there might be no logical response to the existence of recombinant DNA and that we had no recourse but to move ahead. Our civilization had reached its present highly advanced state only by facing the unknown, always hoping that we could generate enough ingenuity to get us out of any jams that might arise.

Instead, the National Academy of Science made what then looked like a prudent response. It put together a high-level committee of scientists who worked with DNA with the hope that they would do more than throw dice that somehow their past experiences would equip them for a logical response. The truth in such situations, however, is often just the opposite. But no one likes to advertise that we may have no meaningful guide for what tomorrow will bring. Psychologically, this is hard to accept, and our sanity almost demands placing more faith in experts than the facts warrant. So when we bring authorities together, we have already half committed ourselves to following their advice. We know that if we don't, we may have to make the decision ourselves and shoulder the criticism if the wrong moves are made.

In particular we mustn't overreact by assuming the worst possible outcome and settling into a siege response that prevents the possibility of progress if in fact nothing catastrophic happens. For example, we would destroy any chances of maintaining our national prosperity if all our governmental actions were predicated upon the belief that the Russians were about to engage us in a nuclear war. And almost equal senseless panic would result if, say, six months of almost no rain led us to half-starve ourselves because of the fear that we were about to enter a decade of no rain.

Likewise, we shall only harm ourselves by assuming that the newly developed recombinant DNA technology poses a credible threat to our civilization. The good that can come out of its use is immense: It can revolutionize our understanding of human chromosomes, give us a practical way to make vaccines against difficult-to-grow viruses like hepatitis, and yield unlimited amounts of now-scarce drugs like the antiviral agent interferon. Nonetheless, there is opposition to its use based on the conjecture that our blessings might indeed be mixed and that recombinant DNA procedures might also generate new disease agents or modify preexisting organisms so as to badly upset the earth's ecology.

Not even hints exist, however, that any of these dour events would be plausible outcomes of totally unrestrained DNA research. Infectious-disease experts tell us that pathogenicity is not easy to generate, and the illegitimate interspecies DNA transfers mediated in nature by viruses have probably already tested the ecological consequence of any DNA transfer that we can now do in the laboratory. We should thus not get into a tizzy because we can't now, or even in the far distant future, have a way to disprove implausible hypotheses about how we may meet our doom. Instead we should push recombinant DNA research and development as fast as our resources permit, while keeping alert to the highly improbable event that one or more lab workers will come down with a disease that we have not seen before.

Unfortunately, we are not now acting with such bold intelligence. We are badly held back by a morass of bureaucratic regulations that are seen by most of the scientists they affect as having no intellectual validity. Alas, I was one of those who helped bring about these increasingly despised rules. Some eleven molecular biologists, of whom I was one, called in the spring of 1974 for a temporary moratorium on experiments involving either tumor virus genomes or the transfer of the genes coding for dangerous toxins from one group of bacteria to another. The moratorium was to last almost a year, until the matter could be discussed by a larger internationally based expert group that was to meet in California at the Asilomar Conference Center. In doing so, we opted for an apparently fair middle position that could satisfy both those many scientists who wanted to move swiftly and those who felt that if we were to err it should be on the side of caution.

This moratorium call, made in half-haste after only a half-day meeting at MIT, was an act I later came to badly regret. The experiments we had so casually halted were soon to strike me as much safer than many categories of work with disease-causing agents that the microbiologists have been carrying out for decades without significant harm to themselves, much less to the public at large.

For example, experiments with tumor virus genes inserted into *E. coli* instinctively appear safer than working with the viruses themselves. Yet there are no firm regulations governing tumor virus research; the precautions to be taken being left to the individual investigator to decide. This seeming indifference to public health is not that at all. Without knowing the level of risk, if any, it is impossible to know what precautions, if any, are necessary. Any rigid rules governing tumor virus research are bound to appear capricious to many, and the only practical recommendation is a prudent respect for commonly accepted microbiological safety procedures.

Why then did we react so much more cautiously to recombinant DNA? A major reason was that the recombinant DNA procedures were not yet a necessary ingredient for our day-to-day research. Until they became so, we did not see the need to appear possibly indifferent to the public good by unilaterally ignoring those who said that science was now out of control. Moreover, Watergate was still with us and the national mood was very much to come clean with what one was up to. So we saw no harm and possibly some considerable credit in so attracting public attention. It was thought best to overestimate rather than underestimate our concerns. Later, when more experiments had been done and no one had taken ill because of recombinant DNA, we could seriously downgrade, if not forget, the whole matter. Then no one could accuse us of keeping back even our most paranoid ideas.

Unfortunately, none of us seriously questioned whether we might be alerting the public unnecessarily and by doing so give recombinant DNA doomsday scenarios a credibility they didn't deserve. The minute the moratorium was announced, we had, in effect, asked the public to join us in the decision-making process. Why would we have actually halted our experiments if we weren't really worried? This point was not initially perceived, nor to my knowledge did anyone see the consequences of forming subcommittees to rate various types of recombinant DNA experiments for their potential risks. If, in fact, risks could be quantitated, they would have to be real, accompanied by the virtual inevitability of the public's insisting that more formal rules take over when the moratorium ended.

The trap was thus almost closed when we assembled at Asilomar in February 1975. By then there had been so much talk to the press and TV about the wisdom of our initial caution that it was generally thought politically unacceptable to argue, as I did, that the whole conference was a mistake and we would do everyone a service by cutting it short with a brief statement to the effect that it was logically impossible to regulate conceptual risks. Already in their opening statements the organizing committee cautioned us not to appear self-serving and instead to come up with guidelines that would have the dual virtue of having teeth yet being imaginative enough so that we would not have to stop too many scientifically valid experiments. This mood was quickly shared by most participants, and speaker after speaker rose to identify his position with that of intelligent caution.

During the next two days there were only a few minor embarrassing outcries that the logic behind the proposed rules was paper-thin if not nonexistent; virtually everyone joined together to come up with a set of internationally applicable rules to govern how everyone should subsequently work with recombinant DNA. In doing so the Asilomar congregants placed their collective reputation behind the statement that the dangers were real enough to embrace four categories of increasing presumptive risk. Some experiments were left almost unregulated while others became saddled with so many precautions that virtually all laboratories would be prevented from their execution.

At first, the Asilomar solution pleased almost every participant except those relatively few who sensed that they had been stopped without a fair hearing and that they had been stymied merely because Asilomar would have been judged a whitewash if no one had been seriously hurt. Most others left for home happy that their own experiments might be done without too much hassle if the Asilomar recommendations were soon accepted by their respective governmental authorities. The general mood was that of relief, and virtually no one faced up to what his response would have been if the majority of experiments had been proscribed as too dangerous for the ordinary lab.

Quickly, however, the uneasy truth began to emerge that Asilomar had been more of an emotional experience than a logical response to firm facts. With no experiments existing that might provide any clues to the magnitude of the theoretical bad, there was no convincing way to defend its conclusions either against those critics, like me, who saw no compelling reasons for any guidelines at all or against the doomsday prophets who wanted all recombinant DNA work to stop. Thus, not unexpectedly, the guidelines began to unravel the moment they became open for formal reexamination.

The initial assault came when the first committee set up by NIH to receive Asilomar met in July 1975 at Woods Hole. It was chaired by Dave Hogness and composed largely of eukaryotic molecular biologists, several of whom by then had already taken up recombinant DNA in a big way. With formal rules now hanging over them, they saw Asilomar as less benign than at first perceived. Even worse, not one of them felt the slightest apprehension about work with recombinant DNA. The only anxiety, in fact, came from the journalistic buildup of Asilomar as one of science's finest hours. Such farfetched exaggeration could only make the public far more concerned than the matter warranted. Hoping to defuse the issue, the Hogness Committee urged a general relaxation of the proposed rules.

In particular they downgraded what they felt were unnecessarily harsh regulations for working with tumor viruses and mammalian DNA.

Almost immediately, however, there were counterarguments to beef up Asilomar from a group of phage workers that met here in Cold Spring Harbor a month later. They proclaimed the Woods Hole action a dangerous example of the failure of self-regulation. Making a virtue of their own immediate lack of need for recombinant DNA technologies, they viewed themselves as more objectively suited to decide which experiments should be carefully regulated. They were particularly bothered by unregulated work with tumor viruses and wanted assurances that they would not needlessly be exposed. However, being totally unqualified to discuss how tumor virus genes inserted into *E. coli* could realistically threaten human beings, they focused mostly on the putative social misuses of recombinant DNA and in particular on how it might hasten the day when we could genetically engineer human life.

NIH quickly caved in and expanded its recombinant DNA Advisory Committee, knowing that they had no logical way of choosing the Hogness Committee's recommendations over those proposed by the self-proclaimed socially conscious phage world. Moreover, with Senator Kennedy taking a populist cry that recombinant DNA was too important to be left in the scientist's hands, they wanted to be protected by a broader-based committee that would listen to all scientific viewpoints including that of the so-called concerned public. Whom from the public to choose, however, was not that obvious, and NIH's first course was to be sure that attention was given to representatives from both the new left, which for political reasons wanted to use recombinant DNA as a rallying cry for the downtrodden in order to get back at MIT, Harvard, and Berkeley, and from the Calvinistic hard core of the environmental movement, which believes that only tighter and tighter regulations could protect us from our baser motives of greed and ambition.

This ploy badly misfired as those who now found recombinant DNA regulation intellectually impossible, and thus absurd, found themselves outnumbered on the official scene. The balance tilted toward tougher regulations, and the final guidelines that emerged from NIH in June of 1976 restricted DNA work far more stringently than those proposed by Asilomar.

Like any compromise between irreconcilable positions, the new guidelines satisfied no one. For example, the environmentalists reacted with lawsuits based on NIH's presumed failure to prepare adequate environmental impact statements. And the Science for the People crowd cried that the guidelines could not be enforced over industry and that the capitalistic imperative for profits would deluge us with newly created deadly germs. Equally angry were the increasingly large number of molecular biologists who could not start their experiments until the certification of the so-called safe strains that none of them felt were at all needed. While so waiting, they could only curse themselves for acquiescing so witlessly in the Asilomar charade. Even worse, they didn't see how they could call off the whole thing without seeming erratically self-serving. Having been greatly praised by the press for their wisdom in calling for Asilomar, virtually none of them was anxious to say he thought that the whole exercise was a silly miscalculation.

Instead, most molecular biologists felt that the best course was to argue in public that the Asilomar guidelines were the correct response and that if we were to follow them we would have nothing to worry about. This tactic, however, frequently created more anxiety than it dispersed, because it implied that we knew what we were protecting ourselves against. If that were so, the dangers that society faced were real and we must worry what would happen if the NIH guidelines are not strictly adhered to. Inevitably there was more and more talk about the need for some form of punitive national legislation, and drafts of proposed laws began to circulate in Congress in 1977.

Initially the virtual certainty of some form of national legislation was accepted by most molecular biologists, who at the first public hearings before Congress almost without exception said that some form of guidelines, if not laws, would be a good thing. This way they hoped not to antagonize those in power and thus hopefully tone down the final harm. Happily, Congress got tangled up in trying to decide which form of legislation was needed, in particular whether it should forestall the passage of even tougher rules by state and local governments. This gave us hope that if we lobbied hard enough the absurdity of recombinant DNA regulation would become clear and the guidelines would gradually fade away before anyone codified them into real law.

In fact this has partly happened, though not as fast as would have occurred if we all quickly said in public what we were endlessly boring ourselves with in private. In any case, the about-face last spring of the

politically astute Senator Kennedy almost certainly signifies that recombinant DNA no longer has any political value and that public hysteria cannot be maintained indefinitely in the absence of a credible villain. To be
sure, not all senators are yet on our side, and Senator Stevenson just
recently issued a lengthy report arguing for legislation. Fortunately his colleague on the Science Committee, Senator Harrison Schmidt, has issued a
strong counterreport and the odds now appear to be against any recombinant DNA legislation going through Congress in the foreseeable future.

Our main concern now is the speed with which NIH can rid itself of
its regulatory role over DNA research. New committees must reverse the
actions of old ones, and approximately one year ago there commenced a
concentrated push to effect a general lowering of all guidelines and in particular to open up tumor virus genomes to recombinant DNA analysis. To
do these latter experiments we then had to go to England or France where
the rules were not nearly as dogmatic as ours. New ad hoc groups thus
were charged by NIH's Director, Don Fredrickson, to come in with arguments for less stringent guidelines. In due course they did their job, and a
by-now somewhat emboldened Recombinant Advisory Committee gave
its consent last June to the desired relaxations. This process, however, was
far from painless. The now absurdly lengthy procedural details that must
be followed to give due process to those who disapprove any downgrading
have consumed a disgracefully disproportinate amount of the NIH direc-
torate's time. However, by late September we thought we were at last
home, with only the remaining step being approval by HEW Secretary Califano. His staff, being lawyers, had no way of independently assessing the
pros and cons, and we initially hoped that Secretary Califano, being a busy
man, would just sign the new guidelines and move on to something to
which he might make a real contribution.

Instead we found that the Secretary's staff was in no hurry to sign and
had homed in on the future complexion of the Recombinant Advisory
Committee, which would have the power to still further downgrade the
regulations. In particular they strongly objected to the proposal of NIH
that the Committee be dominated by leading scientists who wanted to
work with recombinant DNA. Instead they wanted an ethnically balanced
committee that would represent all sides of the matter.

This proposal naturally depressed many of us because it opened up the
prospect of persons without any scientific qualifications deciding what sci-

ence we should not do. Naturally we called on everyone we knew in the White House to alert them to this new folly within HEW. We were told, however, that we shouldn't worry, as in fact the Secretary would on December 15th OK the proposed guideline relaxations, and, moreover, that only a few of the public members would be real obstructionists and we could count on the rest to help us quietly dismantle the whole hateful artifice. Why the White House should see it necessary to put any obstructionists on board at all is beyond me because I doubt that the current administration would lose more than 500 votes over the entire United States were it to announce that we can do anything we want with recombinant DNA.

Now I'm afraid we must remain very vigilant, because the new public members, by being so appointed, naturally presume they have a real job to do. It is not as if they were directors of the Postal Service and getting $25,000 per year for twiddling their thumbs. Instead they all too clearly will be wasting their time if they routinely act as rubber stamps for a scientific community that knows it is hopeless to teach them the subtleties of molecular biology. Their self-respect may thus demand that they try to show that we scientists don't know everything; and so we must expect that they will say "no" more often than we want.

We must not fight back, however, by saying that only we scientists can judge unquantifiable conceptual risks. We are no more qualified than the man in the moon to assess such situations, and we are stewing in this mess because at Asilomar we said we were. On the contrary, if there were firm facts showing us that recombinant DNA research was leading to dangerous new bugs, then it would be totally proper for the public to help us decide what we should do next. Because, however, no data exist to let us decide rationally whether we or the public should worry at all, no Recombinant Advisory Committee should exist at all.

This point, not whether there should be public members, is the crux of the matter. We are upset by the thought of too many public members only because they may take regulation seriously—unlike almost all molecular biologists, who now painfully realize their past blunders and can now be counted on to vote themselves out of existence. But if too many public members exist, they might co-opt enough faint-hearted scientists to delay the dissolution, so that next year at this time, if not for another decade, we shall still be writing up these idiotic memoranda of understanding. We must thus call out as loudly as we can for the ability to do all forms of recombinant DNA research unfettered by any inherently mindless chains.

To act otherwise is against all the traditions that have given us modern science, and we should not think ourselves at all courageous to demand the right to go ahead with new experiments as long as there is no valid reason to presume we might harm someone. Yet Secretary Califano is now proposing that we start serious experiments to show that we shall not be put in danger from our newly created bugs. This is effectively an impossible demand, because some incalculable risks go along with almost all major medical or technological steps, no matter how we proceed. This fact of life cannot be avoided by the setting aside of large sums of money for "risk assessments" that necessarily have no generalized applicability. Given the almost infinite variety of potential recombinant DNA experiments, there is no way that such risk assessments could ever be done on the massive scale necessary to reassure anyone with a real brain. At best the limited highly touted examples would only waste more public monies.

Yet to my amazement Ms. Shirley Williams, the hitherto thought most competent Minister of Education and Science of the British government, has recently stated on TV that they are prepared to spend "tens of millions of pounds" (at least one-third of the entire budget of Britain's Medical Research Council) for recombinant DNA risk assessment. This extraordinary statement is a painful reflection of the gaps that exist between the two cultures, both in England and here in the United States, and we can only hope she is soon put in contact with scientists who have their feet on the ground.

How we are to educate Secretary Califano and his staff is not clear, because they seem to see the political imperative to give almost as much weight to outsiders as they do to the leaders of our profession. Why this should be so is hard to fathom, because the Secretary could not have been indifferent to quality advice in his legal days, and his public views on cigarettes show that he can strongly choose sides.

We thus have no choice as scientists but to make widely known our lack of confidence in Secretary Califano's office, and the possibility that its misguided egalitarianism will deprive the American nation of the full benefits of the most powerful new research tool available for biological research.

The Nobelist Versus the Film Star

1978

Until the last year, I never thought much about my allegiances. My parents were for Roosevelt and against the spoilage of our land by senseless land spectators or industrial giants who put steel mills where there had been sand dunes and the prairie warbler had nested. People who went on bird trips or camped in the national forests and wanted to save Mineral King were the right sort, while those who owned big yachts or stripped the rolling fields of Ohio for coal were the bad guys whom we must get laws to stop. So it was natural to make out a modest check whenever Robert Redford or some equally fine fellow asked you to help him defend the environment and fight the polluters who would give us more cancer.

Now, however, I must confess that I didn't respond to Robert Redford's latest appeal. It is not that I am against him as a folk hero, but, though he must be unaware, he and I are, for practical purposes, real enemies. For some of the money he raises for the Environment Defense Fund is being used to try to stop the experiments we do with "recombinant DNA."

This test-tube-made genetic material now provides an incredibly powerful means to find out what human genes are like. And in so doing it will give us important new ways to think, say, about our immune systems, or how our blood cells are made or the nature of the genes that go out of control when cancer arises.

This being so, I most certainly am a Friend of DNA and want work with recombinant DNA to go as fast as possible. In the old days, this impulse would generally be viewed as good for the earth. Now, however, there exist highly vocal groups who think I'm a danger to the world. The

Friends of the Earth, the Sierra Club, and the Natural Resources Defense Council, as well as the Environmental Defense Fund, all say that our experiments pose a realistic threat to our way of life and must be constrained by their new breed of environmental lawyers.

All this initially surprised me, because I had always regarded environmentalists as among our most intelligent public groups and thought that the original rules for work with recombinant DNA which had come out of the 1975 Asilomar Conference should more than reassure them, particularly because I found those guidelines as a terrible overkill and probably not at all necessary.

My fellow DNA workers wanted, however, to act more than clean and certainly to give the impression of being responsible citizens. So they suggested that we largely work with specifically enfeebled organisms that would not live well outside our test tubes. And when, after Asilomar, the matter was taken up by the National Institutes of Health, they in turn wanted to look like the perfect guardian of our health, and so the guidelines that we now have to live with became more than tough. In fact, they effectively blocked most of the better experiments that directly relate to cancer.

As a result, the DNA community is now very unhappy working under harsh rules we do not believe necessary and which waste vast sums of sorely needed research funds. We now want to relax greatly the guidelines we imposed upon ourselves.

Unfortunately, we find this task to be much more complicated than their original drafting. Our main problem is that in our original statements about recombinant DNA, we kept referring to "potential dangers." Instead we should have said "conjectural dangers," because there was, and still is, not a trace of evidence that any of the experiments pose a threat to those who do them, much less to the general public.

In being so linguistically sloppy, we gave a long awaited opening to the two groups that were out to embarrass us. The first consists of disgruntled long-out-of-productive-science biochemists, who use any opportunity to say bad things about how the effects of modern science are carried out. The other is a tiny, though noisy, group of Boston-based academic leftists who fantasize that the rich will finally subjugate the masses by giving them bad genes manufactured by recombinant DNA methodologies. This is a mad idea that I suspect they are too intelligent really to believe. It must be a tactical move in their zany campaign to convince the Boston poor to rise up against the elitist imperialism of MIT and Harvard.

We never expected, however, that we would be branded as polluters by the environmental movement, because until recombinant DNA came along, we always thought we were on their side.

After all, who wants to see our planet not fit for our children to inherit? When they went to court to block DDT or keep the skies of Monument Valley blue, we could only applaud. So why now are we on opposite sides? Can we have on blinders, and can our self-interest as scientists not allow us to see how indifferent we are to the harm we may do? Might, in fact, the professional environmentalists present arguments that we just can't face up to?

I feel strongly that this is not the case. Compared to almost any other object that starts with the letter D, DNA is very safe indeed. Far better to worry about daggers, dynamite, dogs, dieldrin, dioxin, or drunken drivers than to draw up Rube Goldberg schemes on how our laboratory-made DNA will lead to the extinction of the human race.

The strains of viruses and cells we work with in the laboratory generally are not pathogenic for humans, and all we know about infectious diseases makes it unlikely that the addition of a little foreign DNA will create any danger for those who work with recombinant DNA-bearing bacteria. Even if no special guidelines existed and we only employed the standard microbiological practices of routine sterilization, we should have no reason to be concerned about our health. Equally important, we should not worry that our experiments will profoundly alter evolution by creating bizarre life forms unlike any seen before. DNA is frequently carried from one species to another by viruses, and the global evolutionary impact of our experiments must be negligible compared to naturally occurring DNA transfers.

If this is so, how can we explain the enthusiasm with which so many professional environmentalists wish to shut us down?

The answer, I fear, is that such groups thrive on bad news, and the more the public worries about the environment, the more likely we are to keep providing them with the funds they need to keep their organizations growing. So if they do not watch themselves, they will always opt for the worst possible scenario.

For the short term this may have given them more recruits, but I worry about the long-term effect. No one will benefit if we perceive the credibility of our environmental movements to be no better than that of the most troglodytic of our industrial firms.

If what they say about DNA is nonsense, do we have any compelling reason to listen to them when they come out against pesticides that give us shiny apples or tell us that the waters of the Mississippi are likely to give us cancer? I would like someone to set me right on such matters, but whom to trust now is not that clear.

The DNA Biohazard Canard

1979

The recent relaxation of the NIH guidelines governing recombinant DNA research is an important event. For the first time, it permits any molecular biologist, who so desires, to do recombinant DNA experiments involving mammalian or tumor virus DNA. Under the old guidelines, such work with tumor virus DNA was limited to germ warfare (P4)-type labs and therefore out of reach of most serious scientists. While recombinant research with, say, mouse DNA per se was not so heavily restricted, its limitation to labs (P3) under negative air pressure again meant that many good scientists, especially in the less affluent universities, could not start up such research.

Now all that is necessary for most experiments are (1) the following of standard microbiological procedure in a lab (P2) equipped for work with pathogens, (2) the use of the "so-called safe bacteria and viruses" as cloning vehicles for the recombinant sequences, (3) the submission of detailed experimental protocols to an institutional biohazard committee that includes public members and which must ascertain that the experiments to be done are on the list approved for P2 labs, and (4) some form of regular health surveillance program for the respective scientific and technical staff. If there is ambiguity in the guidelines as to where a proposed experiment falls, then the matter must go to the recombinant DNA division within NIH with the hope that they will say yes to the proposed experiment.

This state of affairs is a vast improvement over the stifling inaction of the past four years even though several classes of potentially important experiments remain effectively proscribed. So no matter how we now feel about the legitimacy of the guideline process itself, perhaps we should happily accept the new rules. Our experiments of the next few years should be most exciting, and more reward should come from concentrating on them as opposed to efforts to still downgrade further the regulations.

I do not think, however, that we should take this breather. To start with, the current regulations necessarily waste money that is best spent on science itself and not on the mindless time-consuming paper and committee work that our current rules still demand. Also, while it sounds fine to insist that the public be informed about what we are doing, would it make sense, say, to take a local doctor away from his patients so that he can learn what we are up to? Only if the arguments behind our current guidelines have any rational basis. Such is not the case; the guidelines are based on the assumption that recombinant DNA is potentially dangerous. However, in the absence of any data as to whether recombinant DNA poses any potential danger at all, it is as impossible for the highly literate layman as for the deepest illiterate to generate an appropriate response.

I and many of my friends thus see the current guidelines largely as a means of proclaiming that we want to be safe rather than as a reflective instrument for advancing either our personal or national healths. If so, our biosafety committee thus functions to give the facade, as opposed to the reality, of effective public concern. Because we believe that its members have no chance to do anything beneficial either to us or to our community, we should go out of our way not to appoint any total outsiders who have alternative activities that will better promote the public interest. Hence our reluctance to take a medic away from his patients or to take an ecology-oriented citizen away from fighting to preserve our local wetlands.

This leads to the question of who are the two outsiders who have joined our biosafety committee so that we can proceed without loss of our government support. One is a scientist at the Brookhaven National Laboratory who, like us, works with adenoviruses. He already knows all the rules, and so he need not waste more time poring over the regulations. The other is a local conservationally minded artist, well known in our community and who once studied English at Smith. She lives only a few minutes' walk from our labs; and because she does not have a nine-to-five job, she is available at short notice as opposed to those with regular jobs for whom we would have to schedule evening or weekend meetings. That she knows virtually no science is an asset, because she will not wrestle over our protocols with the thought that if only she learned a few more terms she would better further the public interest.

I believe that the total cynicism with which we here in Cold Spring Harbor now regard the whole regulatory process is shared worldwide by virtually all the labs that now work with recombinant DNA. The filling out of still another memorandum of understanding and agreement (MUA) strikes us as this decade's equivalent of the anti-Communists affidavits that marked the university scene in the late 1940s or the Declaration to the Flag that we made in still earlier epochs. Then we had to proclaim our virtue by saying we were against Communism and for the American flag. Now we are required to say that we are for safety even though we have not the foggiest idea whether any of the prescribed rules will give us even an ounce of safety. This latter point is not trivial. As the physicist Freeman Dyson has recently so aptly stated, the price of saying no may be far greater than that of saying yes. For example, by still being effectively blocked from work with tetanus toxin, we may be depriving ourselves of a powerful means to learn how it functions. I shudder at the thought that our future might increasingly be in the hands of habitual pessimists who instinctively predict anything new will make us worse off.

It furthermore may be very difficult to maintain effective compliance over our now only modestly annoying regulations for any extended period of time. Policing each experimental step is effectively impossible, even if each of our labs had its own DNA policeman, and when no one is about I suspect that many shortcuts will be taken. It is not that we as scientists are more immoral than any other group. It is because we are no different from the rest that I fear we shall not follow into perpetuity rules (e.g., the wearing of lab coats, the signing of log books, the closing of doors, the banning of Tab, etc.) that we do not think will make us any safer. In obeying orders that are there only to look good, we are bound to deemphasize real dangers like radioactivity or ether explosions, and so long as our Safety Officer thinks he must take recombinant DNA regulation seriously, he will not be overseeing steps that might avert well-characterized lab accidents.

The continuation of the guidelines furthermore can only lead to a diminution of NIH's reputation among the scientists it serves. Every time we receive still another heavy packet of the Recombinant DNA office's talmudic-like legalese, we necessarily question the competence of those who either write or sign them. Though we want to believe that Don Fredrickson knows he has become a massive signer of junk mail, how can we be sure when these senseless directives give us a slight hint of what it must be like to be a Russian scientist.

There is also the question of how to behave in face of the fact that different countries have different rules, and that an experiment effectively forbidden here might be permitted in England or France. Until our guidelines were lowered, this was, in fact, the case with work on tumor viruses, and several of our staff spent part of last year in London. Now the tables are reversed, and unless the British Genetic Manipulation Advisory Group (GMAG) bureaucrats back off, the future of British virology over the next decade is indeed bleak. Naturally, their best younger virologists now want to come here, and until several days ago we were prepared to receive one such scientist. As of several days ago, however, he had been unable to obtain permission from their Medical Research Council, a decision that apparently was taken at the ministerial level during England's latest case of crippling strikes. And we Americans are now also told that we can no longer go abroad to seek out sites where we can more efficiently carry out our NIH-suppressed ideas. Whether any of us in frustration will mimic the Hollywood Ten and do our science under assumed names is not clear, but as long as we feel such total contempt for the regulations, the possibility is bound to arise.

We must thus continue to lobby vigorously to bring this miserable canard to its fitful end. Already many of the major legislative figures in this country (though not in England) are coming to realize that rationally regulating recombinant DNA research has the same chance of success as the prevention of illegitimate sex. There do remain minor demagogues who like the sound of legislative bans against test-tube genes, but they should get nowhere if we molecular biologists effectively band together to argue our cause.

Much more important will be the actions of the Recombinant Advisory Committee (RAC), which now has the power to recommend to HEW the virtual disappearance of all guidelines. Recently it has effectively been taken away from NIH and placed under the HEW Secretary Califano. Not liking the sound of a scientist overdominated committee, he has more than doubled its previous size (11) by the inclusion of 13 new members, the vast majority having no connection with DNA research. To say the least, the recombinant DNA community is nervous, because the new RAC members, being largely a representative sample of the Democratic party, could be a somewhat unpredictable mix of sharp brains, environmental overreactors, intelligent goodwill, and leftish kookery.

Conceivably the new RAC may surprise us and quickly vote to restore the scientific freedom that has served us so well in the past. I suspect, however, that most members, having no personal desire to do such work themselves, will see no rush to take this apparently drastic step that would take them out of existence. They are likely, at least initially, to be more influenced by whether a proposed experiment sounds like a menace to public health (e.g., work with the genes for botulinus toxin) than by the nonexistence of any facts that should give us legitimate cause for concern. What worries us most is the proposal that we scientists will be asked to prove what we want to do is safe, as opposed to a carte blanche to go ahead unless there is demonstrable evidence of potential harm. While at first glance this may sound like a legitimate demand from a population already edgy about nuclear bombs, there is no way that such assurances can be given about novel experiments not yet done.

Our particular plight must be seen as part of a never-ending battle for our freedom to talk and act as we wish as long as our actions do not threaten the lives of others. We must always be prepared that others will get nervous at the prospect of major innovations like recombinant DNA and will try to block them. But when they cry stop, we must not fall into the bottomless trap of, say, having to logically disprove the future occurrence of monsters. This is an impossible task. Rather it is those who want recombinant DNA regulation who must make a credible case that they could arise. Otherwise, we may gradually move into a situation where not only our DNA research, but that of all scientists, is increasingly regarded as fair game for capricious bureaucratic interference.

Thus we must know what we want when we say that the science is too important to be out of public control, because control, without the capacity to act responsibly, is at best chaos and at worst tyranny. As responsible citizens we must be very careful to keep away from issues like recombinant DNA regulation where we can talk forever but never think. This is easier said than done, but I am still enough of an optimist to believe that common sense must eventually prevail.

Ethos of Science

Moving Toward the Clonal Man:
Is This What We Want?

1972

The notion that man might sometime soon be reproduced asexually upsets many people. The main public effect of the remarkable clonal frog produced some ten years ago in Oxford by the zoologist John Gurdon has not been awe of the elegant scientific implication of this frog's existence, but fear that a similar experiment might someday be done with human cells. Until recently, however, this foreboding has seemed more like a science fiction scenario than a real problem that the human race has to live with.

The embryological development of man does not occur free in the placid environment of a fresh-water pond, in which a frog's eggs normally turn into tadpoles and then into mature frogs. Instead, the crucial steps in human embryology always occur in the highly inaccessible womb of a human female. There the growing fetus enlarges unseen, and effectively out of range of almost any manipulation except that which is deliberately designed to abort its existence. As long as all humans develop in this manner, there is no way to take the various steps necessary to insert an adult diploid nucleus from a preexisting human into a human egg whose maternal genetic material has previously been removed. Given the continuation of the normal processes of conception and development, the idea that we might have a world populated by people whose genetic material was identical to that of previously existing people can belong only to the domain of the novelist or moviemaker, not to that of pragmatic scientists who must think only about things that can happen.

Today, however, we must face up to the fact that the unexpectedly rapid progress of R.G. Edwards and P.S. Steptoe in working out the conditions for routine test-tube conception of human eggs means that human embryological development need no longer be a process shrouded in secrecy. It can become instead an event wide open to a variety of experimental manipulations. Already the two scientists have developed many embryos to the eight-cell stage, and a few more into blastocysts, the stage where successful implantation into a human uterus should not be too difficult to achieve. In fact, Edwards and Steptoe hope to accomplish implantation and subsequent growth into a normal baby within the coming year.

The question naturally arises, Why should any woman willingly submit to the laparoscopy operation that yields the eggs to be used in test-tube conceptions? There is clearly some danger involved every time Steptoe operates. Nonetheless, he and Edwards believe that the risks are more than counterbalanced by the fact that their research may develop methods that could make their patients able to bear children. All their patients, though having normal menstrual cycles, are infertile, many because they have blocked oviducts that prevent passage of eggs into the uterus. If so, in-vitro growth of their eggs up to the blastocyst stage may circumvent infertility, thereby allowing normal childbirth. Moreover, because the sex of a blastocyst is easily determined by chromosomal analysis, such women would have the possibility of deciding whether to give birth to a boy or a girl.

Clearly, if Edwards and Steptoe succeed, their success will be followed up in many other places. The number of such infertile women, while small on a relative percentage basis, is likely to be large on an absolute basis. Within the United States there could be 100,000 or so women who would like a similar chance to have their own babies. At the same time, we must anticipate strong, if not hysterical, reactions from many quarters. The certainty that the ready availability of this medical technique will open up the possibility of hiring out unrelated women to carry a given baby to term is bound to outrage many people, because there is absolutely no reason why the blastocyst need be implanted in the same woman from whom the preovulatory eggs were obtained. Many women with anatomical complications that prohibit successful childbearing might be strongly tempted to find a suitable surrogate. And it is easy to imagine that other women who just don't want the discomforts of pregnancy would also seek this very different form of motherhood. Of even greater concern would be the potentialities for misuse by an inhumane totalitarian government.

Some very hard decisions may soon be upon us. It is not obvious, for example, that the vague potential of abhorrent misuse should weigh more strongly than the unhappiness that thousands of married couples feel when they are unable to have their own children. Different societies are likely to view the matter differently, and it would be surprising if all should come to the same conclusion. We must, therefore, assume that techniques for the in-vitro manipulation of human eggs are likely to become general medical practice, capable of routine performance in many major countries, within some 10–20 years.

The situation would then be ripe for extensive efforts, either legal or illegal, at human cloning. But for such experiments to be successful, techniques would have to be developed which allow the insertion of adult diploid nuclei into human eggs that previously have had their maternal haploid nucleus removed. At first sight, this task is a very tall order because human eggs are much smaller than those of frogs, the only vertebrates that have so far been cloned. Insertion by micropipettes, the device used in the case of the frog, is always likely to damage human eggs irreversibly. Recently, however, the development of simple techniques for fusing animal cells has raised the strong possibility that further refinements of the cell-fusion method will allow the routine introduction of human diploid nuclei into enucleated human eggs. Activation of such eggs to divide to become blastocysts, followed by implantation into suitable uteri, should lead to the development of healthy fetuses and subsequent normal-appearing babies.

The growing up to adulthood of these first clonal humans could be a very startling event, a fact already appreciated by many magazine editors, one of whom commissioned a cover with multiple copies of Ringo Starr, another of whom gave us overblown multiple likenesses of the current sex goddess, Raquel Welch. It takes little imagination to perceive that different people will have highly different fantasies, some perhaps imagining the existence of countless people with the features of Picasso or Frank Sinatra or Walt Frazier or Doris Day. And would monarchs like the Shah of Iran, knowing they might never be able to have a normal male heir, consider the possibility of having a son whose genetic constitution would be identical to their own?

Clearly, even more bizarre possibilities can be thought of, and so we might have expected that many biologists, particularly those whose work impinges upon this possibility, would seriously ponder its implication and begin a dialogue that would educate the world's citizens and offer suggestions that our legislative bodies might consider in framing national science policies. On the whole, however, this has not happened. Though a number of scientific papers devoted to the problem of genetic engineering have casually mentioned that clonal reproduction may someday be with us, the discussion to which I am party has been so vague and devoid of meaningful time estimates as to be virtually soporific.

Does this effective silence imply a conspiracy to keep the general public unaware of a potential threat to their basic ways of life? Could it be motivated by fear that the general reaction will be a further damning of all science, thereby decreasing even more the limited money available for pure research? Or does it merely tell us that most scientists do live such an ivory-tower existence that they are capable of thinking rationally only about pure science, dismissing more practical matters as subjects for the lawyers, students, clergy, and politicians to address?

One or both of these possibilities may explain why more scientists have not taken cloning before the public. The main reason, I suspect, is that the prospect to most biologists still looks too remote and chancy—that is, not worthy of immediate attention when other matters, like nuclear-weapon overproliferation and pesticide and auto-exhaust pollution, present society with immediate threats to its orderly continuation. Though scientists as a group form the most future-oriented of all professions, there are few of us who concentrate on events unlikely to become reality within the next decade or two.

To almost all the intellectually most adventurous geneticists, the seemingly distant time when cloning might first occur is more to the point than its far-reaching implication, were it to be practiced seriously. For example, Stanford's celebrated geneticist, Joshua Lederberg, among the first to talk about cloning as a practical matter, now seems bored with further talk, implying that we should channel our limited influence as public citizens to the prevention of the wide-scale, irreversible damage to our genetic material that is now occurring through increasing exposure to man-created mutagenic compounds. To him, serious talk about cloning is essentially crying wolf when a tiger is already inside the walls.

This position, however, fails to allow for what I believe will be a frenetic rush to do experimental manipulation with human eggs once they have become a readily available commodity. And that is what they will be within several years after Edwards–Steptoe methods lead to the birth of the first healthy baby by a previously infertile woman. Isolated human eggs will be found in hundreds of hospitals, and given the fact that Steptoe's laparoscopy technique frequently yields several eggs from a single woman donor, not all of the eggs so obtained, even if they could be cultured to the blastocyst stage, would ever be reimplanted into female bodies. Most of these excess eggs would likely be used for a variety of valid experimental purposes, many, for example, to perfect the Edwards–Steptoe techniques. Others could be devoted to finding methods for curing certain genetic diseases, conceivably through use of cell-fusion methods that now seem to be the correct route to cloning. The temptation to try cloning itself thus will always be close at hand.

No reason, of course, dictates that such cloning experiments need occur. Most of the medical people capable of such experimentation would probably steer clear of any step that looked as though its real purpose were to clone. But it would be shortsighted to assume that everyone would instinctively recoil from such purposes. Some people may sincerely believe that the world desperately needs many copies of really exceptional people if we are to fight our way out of the ever-increasing computer-mediated complexity that makes our individual brains so frequently inadequate.

Moreover, given the widespread development of the safe clinical procedures for handling human eggs, cloning experiments would not be prohibitively expensive. They need not be restricted to the superpowers. All smaller countries now possess the resources required for eventual success. Furthermore, there need not exist the coercion of a totalitarian state to provide the surrogate mothers. There already are such widespread divergences regarding the sacredness of the act of human reproduction that the boring meaninglessness of the lives of many women would be sufficient cause for their willingness to participate in such experimentation, be it legal or illegal. Thus, if the matter proceeds in its current nondirected fashion, a human being born of clonal reproduction most likely will appear on the earth within the next 20–50 years, and even sooner, if some nation should actively promote the venture.

The first reaction of most people to the arrival of these asexually produced children, I suspect, would be one of despair. The nature of the bond between parents and their children, not to mention everyone's values about the individual's uniqueness, could be changed beyond recognition by a science which they never understood but which until recently appeared to provide more good than harm. Certainly to many people, particularly those with strong religious backgrounds, our most sensible course of action would be to deemphasize all those forms of research that would circumvent the normal sexual reproductive process. If this step were taken, experiments on cell fusion might no longer be supported by federal funds or tax-exempt organizations. Prohibition of such research would most certainly put off the day when diploid nuclei could satisfactorily be inserted into enucleated human eggs. Even more effective would be to take steps quickly to make illegal, or to reaffirm the illegality of, any experimental work with human embryos.

Neither of the prohibitions, however, is likely to take place. In the first place, the cell-fusion technique now offers one of the best avenues for understanding the genetic basis of cancer. Today, all over the world, cancer cells are being fused with normal cells to pinpoint those specific chromosomes responsible for given forms of cancer. In addition, fusion techniques are the basis of many genetic efforts to unravel the biochemistry of diseases like cystic fibrosis or multiple sclerosis. Any attempts now to stop such work using the argument that cloning represents a greater threat than a disease like cancer is likely to be considered irresponsible by virtually anyone able to understand the matter.

Though more people would initially go along with a prohibition of work on human embryos, many may have a change of heart when they ponder the mess that the population explosion poses. The current projections are so horrendous that responsible people are likely to consider the need for more basic embryological facts much more relevant to our self-interest than the not-very-immediate threat of a few clonal men existing some decades ahead. And the potentially militant lobby of infertile couples who see test-tube conception as their only route to the joys of raising children of their own making would carry even more weight. So, scientists like Edwards are likely to get a go-ahead signal even if, almost perversely, the immediate consequences of their "population-money"-supported research will be the production of still more babies.

Complicating any effort at effective legislative guidance is the multiplicity of places where work like that of Edwards could occur, thereby making unlikely the possibility that such manipulations would have the same legal (or illegal) status throughout the world. We must assume that if Edwards and Steptoe produce a really workable method for restoring fertility, large numbers of women will search out those places where it is legal (or possible), just as now they search out places where abortions can be easily obtained.

Thus, all nations formulating policies to handle the implications of in vitro human embryo experimentation must realize that the problem is essentially an international one. Even if one or more countries should stop such research, their action could effectively be neutralized by the response of a neighboring country. This most disconcerting impotence also holds for the United States. If our congressional representatives, upon learning where the matter now stands, should decide that they want none of it and pass very strict laws against human embryo experimentation, their action would not seriously set back the current scientific and medical momentum that brings us close to the possibility of surrogate mothers, if not human clonal reproduction. This is because the relevant experiments are being done not in the United States, but largely in England. That is partly a matter of chance, but also a consequence of the advanced state of English cell biology, which in certain areas is far more adventurous and imaginative than its American counterpart. There is no American university that has the strength in experimental embryology that Oxford possesses.

We must not assume, however, that today the important decisions lie only before the British government. Very soon we must anticipate that a number of biologists and clinicians of other countries, sensing the potential excitement, will move into this area of science. So even if the current English effort were stifled, similar experimentation could soon begin elsewhere. Thus it appears to me most desirable that as many people as possible be informed about the new ways of human reproduction and their potential consequences, both good and bad.

This is a matter far too important to be left solely in the hands of the scientific and medical communities. The belief that surrogate mothers and clonal babies are inevitable because science always moves forward, an attitude expressed to me recently by a scientific colleague, represents a form of laissez-faire nonsense dismally reminiscent of the creed that American business, if left to itself, will solve everybody's problems. Just as the success of a corporate body in making money need not set the human condition ahead, neither does every scientific advance automatically make our lives more "meaningful." No doubt the person whose experimental skill will eventually bring forth a clonal baby will be given wide notoriety.

But the child who grows up knowing that the world wants another Picasso may view his creator in a different light.

I would thus hope that over the next decade wide-reaching discussion would occur, at the informal as well as formal legislative level, about the manifold problems that are bound to arise if test-tube conception becomes a common occurrence. A blanket declaration of the worldwide illegality of human cloning might be one result of a serious effort to ask the world in which direction it wished to move. Admittedly the vast effort, required for even the most limited international arrangement, will turn off some people, namely, those who believe that the matter is of marginal importance now and that it is a red herring designed to take our minds off our callous attitudes toward war, poverty, and racial prejudice. But if we do not think about it now, the possibility of our having a free choice will one day suddenly be gone.

The Dissemination of Unpublished Information

1973

As a boy in Chicago, I used to equate good manners with wearing rubbers over your shoes, something your parents got mad about when you forgot, but somehow sissy and not at all connected to the world of science that I wanted to be part of. Then I only worried about the limitations of my brain and took comfort that no Emily Post-type manual was about to restrict my future actions. I regarded eccentricity, if not total unpredictability, to be the mark of the free mind, with Rex Harrison's portrayal of Henry Higgins pointing the way toward true success. Conventions remained for minor minds unwilling to take chances, and so why pay lip service to forms of behavior that at best generate more hypocrisy?

In my more-than-awkward moments, however, I would get scared that later I would get hell from people who mattered, but never to the point of learning when to wear a tie or to avoid talking about Roosevelt with my father's Republican relatives. Accepting nonsense to please your elders would inevitably lead to similar compromises in later life. So I had little to talk about with most neighborhood kids, already brainwashed into respectability, and even in college I could communicate only with apparent oddballs equally uninterested in ordinary behavior.

Happily, my first contact with high-power minds, which came in 1947, soon after I went to Bloomington as a graduate student of the geneticist S.E. Luria, confirmed my adolescent fantasies. The truly bright did not live like our relatives or nearby neighbors and wasted little time worrying about how they looked, or the polish on their cars, or whether their lawns were overrun with crab grass. They had the guts to concentrate on the

91

important, knowing that they must banish the pedantic if they were to come up with some new law of nature. And if they seemed contemptuous of those stuck in the recent past, their intolerance was in good cause: no one benefits from false praise, and only if the truth is honestly faced does the possibility exist for rebirth. I thus had no doubt that I had become part of the highest form of human achievement, light-years away from the uninformed prejudices of the poor or the callous self-satisfaction of the educated rich.

Then everyone important in modern genetics was constantly in touch, in the winter by post or by going to informal meetings, while the long summer periods were for planning joint experiments when climbing in the mountains or lying on the beaches. As long as you thought about the gene while eating and drinking, there were no age barriers to keep you from the famous, and if they talked nonsense you did not abort your career by setting them right; just the opposite, since persistent politeness was easily mistaken for vapidity, and if you didn't speak your mind who would know that you could hustle and pounce upon clues that others had missed. To be sure, there were those whose facts frequently faltered, but never trying to make it was a much deeper vice than occasional self-deception.

Top-flight genetics thus never attracted a homogeneous congregation, and nothing so surely failed as attempts to whip its recalcitrants into line. Inconsiderate actions, though too common to be unexpected, were naturally not welcomed, and any time might see one or two outcasts designated as beyond the pale. But if they had new ideas or experiments to report, they eventually slipped back and the nature of their black deeds faded when new forms of arrogance were more than you could take. The essence of the gene, however, demanded bold ways of thinking, and life was too precious to waste worrying too long about how your colleagues behaved.

Now, some twenty-five years later, I often have the sense that I still belong to that half-mad, but uniquely wonderful, playing field of my youth, where the aim was the truth, not money, and where decency always took precedence over cunning. But now I know the matter to be more complex, with success in generating new ideas usually being more than the simple combination of native intelligence and good measure of luck. All too often, science resembles playing poker for very high stakes, where revealing one's hand prematurely only makes sense when you have nothing to show.

The basic dilemma is that there are never enough good new ideas to go around. If most scientists were content to take joy in the work of others and not be judged by their ability to generate new facts or ideas, there would be no problem. But most of the scientists I know chose their profession because they were fascinated by the knowable unknown, always hoping that they could be part of the conquering body. To be sure, we often take joy in the discoveries of others, but often only in proportion to the extent that we were not close to the same objective. Mountains are seldom known by their second ascent even when the ways to the top took very different routes. Likewise, credit in science almost always goes only to the person who first saw what was up.

This constant preoccupation with priority is not merely a matter of pride. If it were, we should worry that we are constantly abetting infantile behavior that takes away much of the inherent joy of science, giving to it characteristics that dominate the rat race of the business world. But we all know too well that the types of jobs we eventually get are very much dependent upon how much we produce. There is little enthusiasm for those who always come in second. The key words are productivity and originality, and if those are lacking, the prospects are poor of finding jobs and research money that will let you continue to do high-powered science. Correspondingly, if one hits the jackpot, then he is assured of an interesting position together with piles of money to generate new data.

Now, alas, I can see no chance that this inherently stressful situation will ever change. All signs point toward a future where our inherent wealth remains grossly inadequate for our basic social needs, much less for the total flowering of the intellectual life. Thus, science is never likely to be supported anywhere near the scale that will satisfy its performers, and only a small fraction will ever have the resources they would need for the full expression of their talents.

Thus, not only the self-respect but also the way of life of a scientist depends upon his relative ranking among his peers. As long as he retains the ambition of his youth, he is almost always hoping that he may soon pull off something smashingly big that will intrigue his colleagues. And, giving him the encouragement that will tempt him back to the lab at night is the fact that the time span of a real discovery can be very short, maybe only a few hours to a week or so. Though many years of patient work might be necessary to prepare for the crucial experiment, the more relevant fact may be the brevity of the moment of revelation. One moment nothing, the next everything, or at least until euphoria has worn off and you become stumped by some new puzzle.

Usually you discover that you are not the only person working on a given problem. Only when working with trivia can you be sure that you will be alone. In any given branch of science, there usually exist a series of well-defined objectives, which if achieved should explain vast areas of natural phenomena. Everyone in a given area knows them, frequently ranking them in the same order of importance. Very often, however, the top goal is recognized as insurmountable today, maybe having to wait one to several decades before enough groundwork can be laid for the final attack. If so, lesser targets are picked, usually by some form of calculation that combines your inherent curiosity with your chance to get there first. No good can come following a path already well-traversed, and most certainly never pick a problem that you feel has a very good chance of being solved by someone else. At best, you would be considered derivative, and at worst you could be ostracized by your peers for poaching.

Unfortunately, in most cases this seemingly black-and-white advice is easier to state than to follow. Because science deals with the unknown, the steps to the top are never clearly laid out, and more than one approach may be necessary before the right one becomes apparent. And because each may require the mounting of a substantial effort, any one person or lab usually does not have the resources to move ahead on all fronts. So when the choice is made, it may not be easy later to change your mind, even when you have suspicions you are facing a very steep wall. Equipment has been bought, assistants hired for quite specific tasks, and enough progress made to make you suspect that, given enough time, you will make it. But, all too often, you then become aware that someone else has the same objective but is using a new approach that might make more sense than yours.

Generally, the first reaction to the prospect of being scooped is a combination of despair and hope that your opponent, "X," will fall dead. You may consider giving up, but this could leave you without any tangible results to show for years of toil. Furthermore, you often don't have other goals that really strike your fancy. So you may just plow ahead, fearing someday you will see "X" grinning down from the top. So it is hard not to think about retooling your effort to try the same approach as your competition. Even though you are behind, by being a little more clever you might overtake him. He, of course, might then become hellishly mad, believing correctly that your success is a direct consequence of his having discovered the right way to proceed. So you can be almost certain of making a long-time enemy, the more so depending on the importance of the final breakthrough.

Often the morality of such situations is judged by how you first realized that "X" found a better approach to your problem. If you learned it from an article in a scientific journal in which "X" published details of his approach, then most people will feel that the problem is up for grabs. But if you learned of his trick by hearing him speak informally at a meeting, then you might figure you should let him exploit his observations until his first publication. But if what you have learned totally nullifies your current experiments, you could be driven batty sitting on your hands until your colleague writes up his data, particularly if you figured that he was delaying publication until he had everything sewn up.

Even worse, there can be up to a year-long gap between when a manuscript is submitted for publication and when it finally is printed. In this interval, the manuscript is frequently sent by the editor to referees to see whether it deserves publication. Usually, the best referees are those who work in similar areas and therefore exactly those people most likely to be in conflict of interest with the data they read. While there is often the feeling that referees should refrain from taking advantage of their position, and not tell their students or colleagues juicy new facts, only an insensitive fool could let his student go on with an experiment when his "insider" information tells him that it no longer makes sense.

Moreover, most manuscripts, even before acceptance, are duplicated, with copies (preprints) given to students in the respective lab(s) and to selected colleagues throughout the world. So, in most cases, many interested people besides the referees "legitimately" know of key results long before they are published. Thus, my feeling is that what counts is not when some new result is finally published, but when its discoverer sees fit to write it up in manuscript form, because then, unless he is obsessively secretive, his news soon will be dispersed by the verbal grapevine throughout the interested scientific community. Of course, the cautionary refrain is often heard, don't tell so-and-so because he is a thief, but such warnings seldom work, either because someone has passed the news without the enjoining injunction or because the concept of secrecy quickly loses its force when the originator of the news is no longer directly involved in its spread. On the whole, I find it impossible to keep "secrets" unless they are totally without interest, and so I forget them, or because so-and-so's reputation for piracy just can't be avoided.

Essential to the maintaining of some fabric of decency is the unambiguous attribution of the receipt of unpublished information, for example by the direct statement at the start of a paper that the respective experiments "were done with the knowledge of another person's results." This may not always mollify "X," but failure to do so and to write up derivative work as if it were totally original will bring forth cries of murder. But, all too often, attributions of credit are so backhanded that only "insiders" know whether two identical findings were independent events or whether one reflected work only started upon the receipt of a hot rumor. In particular, many latecomers feverishly rush their claims into quick publication so that they can have the same calendar year as the original observation. While, at first, most interested spectators know the copy, with time such unwritten gossip fades and in later years equal credit is frequently extended.

Unhappily, the apparent quality of the science does not always mark the person who saw the truth first. It is not only second-rate scientists who behave improperly, but often those of first rank. Major success early in life all too easily creates the appetite for more of the same, and many of our best scientists compulsively regard all relevant new discoveries as natural objects for exploitation by their own research groups. They know all too clearly that well-financed research empires depend on the constant manufacture of important new data. While there exist many scientific giants who need no help in generating one important new fact after another, we must anticipate that the prospect of loss of face, if not of empire, will subconciously lead many generally honorable men to actions they will later try to forget.

So only the uninformed or the forever naïve expect science to be a totally open discipline whose participants always freely discuss their newest findings. While I know many scientists who instantly blurt out all they know, despising the duplicity that can go with keeping key information back, they may not be in the majority. All of us know enough horror stories to realize that some scientists are best avoided. Even if you don't care whether they try to move in on your own baby, there are times when you must defend your students from their predatory claws. So I have friends who urge their students to lie low when a pirate is on the horizon, knowing all too well that their youthful enthusiasm may become masked with cynicism if their hard-earned results suddenly roll out of another laboratory.

There is no need, however, to debate the merits of secrecy when a discovery is so all-embracing that it sews up a field. Then you have nothing to lose by immediately shouting it to the world. The double helix that Francis Crick and I came up with in the spring of 1953 was such a case. So naturally we broadcast it as fast as possible, knowing that if we would wait, someone else would inevitably think out the right answer and we would have to share credit. Desires to ward off competition, however, may also tempt you to make claims beyond your data, with the thought that others will give up the chase and for a few months let you do science at a more civilized pace. Assertions are often made informally that so-and-so has been tightly established, say for example the purification of a key enzyme then at the center of biochemistry. Receipt of such news might induce you to give up a similar project, because it would be criminal to keep a younger colleague on a project that already has been accomplished. Only months later do you realize that the key experiments had never been done correctly and that the problem is still for picking. So a discerning eye is often needed to know which bits of unpublished information should be taken seriously.

A clear head is also needed when you hit upon a crucial technique or observation that immediately will make possible scores of other important experiments. If you already possess a large lab with many competent students, then you can have the joy of following up your discovery to its natural conclusion. But if you are unknown and with little resources at your disposal, you may be certain that if you talk openly, other labs will rush in and you may quickly be frozen out of a field that owes its existence to your ingenuity. The only sure way to prevent such a disaster is to let no one outside your lab onto your secret until you have exploited all its obvious consequences. Then you can have the pleasure of a public announcement, uncomplicated by qualms that you are creating unwanted competition. Of course, it may be very tricky not to let your exuberance betray that you have struck oil, and rumors may fly that you have a big secret. Some people may be upset, but usually only because they figure that you must have let your close friends in on your act and they feel discriminated against. Once, however, you reveal what is up, the beauty of your results will soon make most everyone forget your long period of silence.

In contrast, your reputation is likely to suffer if you make a stink about unwanted competition. Your original generosity in letting everyone onto your big secret early could easily be forgotten if you want to eat your cake and then have it. There is no unwritten law among scientists that given problems belong to specific individuals, and agreements among peers on how to divide a tasty pie usually lead to further anguish when it becomes apparent that one side has got the better bargain. To be sure, special pleading may keep some compassionate rival from taking up the scent for a few months, but if what you have announced is at all revolutionary, you must expect the deluge, if not from one side, most certainly from the other. So once your news is out, the only sane course to follow is to work like the devil, hoping that your incipient competition may take several months to get into the act.

Aside from deferences owing to close friendships, which can extend over national boundaries, the only rule that most people follow is the avoidance of problems that someone in their own environment has already latched onto. Only saint-like minds can watch someone in the next lab race them for an experimental result and not get violently upset. Even when minor scientific points are involved, someone's temper will give way, and if friendship ever existed, it will soon turn to antipathy, if not hatred. While competition hundreds of miles away need not upset your day-by-day existence, when it is next door it becomes a canker the brain cannot ignore. This hard fact of scientific life means you may be in for real torture if someone in your own university, if not department, makes an unexpected observation that you would like to exploit. You know that if you were at another university, you would have no hesitation in jumping in on the fun, but if you must coexist within the same faculty, you should tolerantly grin and hope that after the first flush of excitement passes, there may be an oblique way to get into the action.

To be sure, I know of people who have not followed that advice, usually on the false assumption that they have been begotten with talents not given to those who started the ball rolling. However, such bully-like behavior is difficult to pull off, and even when the most insensitive of egotists move in, the storms they ignite are usually more than they have bargained for. If not stopped, they can often paralyze an institution, and those at the top have no recourse but to tell the offending characters to back down or get out. If cooperation instead of competition were to emerge, everyone would breathe a sigh of relief, but if someone wants to do his own thing, he should have his way. Of course, there are many cases when a lone wolf would have been very wise to get more talent on his side and suffered badly for trying experiments over his head. Much depends upon how help is offered, because no one ever feels happy that his own small operation has become part of a large corporation over which he has no effective control. Owning shares in a conglomerate may make sense if you sell out at the right time, but all too often they are not worth the paper they have been printed upon.

Always a key question is the number of people who optimally should be in a field at a given time. All of us would be very bored if we did not have the results of others to look forward to, even within our own highly specialized area. I know of no one who can generate anywhere near enough unexpected new facts to provide internal contentment. New issues of our scientific journals are all too often needed to knock us out of the lethargic grip of our own thoughts. And though we may occasionally complain that there are too many scientific meetings, they are generally indispensable for providing advance clues as to what will come next. Your pulse, of course, begins to skip when you see a title that might imply a solution to years of your research life, and what a relief it is to generally discover that someone else also does not know what is up.

So sanity generally demands some undefinable balance between your productivity and that of the outside. As long as you can continue to turn out good ideas, you can take joy in the success of others and actually promote their careers as essential to your own. But if you are going through lean years and see your research support dwindling, the cheerful arrogance that characterizes tough science on the move may look increasingly indistinguishable from the ethic of the business scoundrels who worry about human frailty only as it affects profits. The love affair with the scientific life that so dominates our lives as young adults is not easy to maintain and, like a real marriage, will most certainly collapse if you expect too much of your love object. Unfortunately, the treatment of the scientific life fed to us as children bears little relation to reality, and the glowing portraits of our selfless characters that dominate virtually all popular books about science read as if cribbed from the lives of the unmartyred saints. So the fact that scientists are no better or worse than others of our backgrounds often is not learned in time to prevent much unnecessary disillusionment.

Care especially must be taken not to credit scientists whose research bears directly on human problems as less liable to human temptation than those who fix on areas lacking immediate relevance to mankind. The press releases that tell us of vast collaborative efforts to understand horrific diseases or to bring forth necessary new forms of energy all too easily can be read to mean that we are witnessing the births of new forms of human cooperation that will light the way for future decades. My experience with such programs, however, is that this is all hogwash. Such programs often get going not because they make scientific sense but because the public is crying for relief and politicians do not like to admit they are impotent to help their constituents. So, independent of whether we have a fighting chance, we can all too easily charge ahead, say against the boll weevil, and in the long term only make some cotton congressman happy, by providing soft jobs for his less accomplished relatives.

As a result, scientists who, for better or worse, get involved in such applied projects usually have poorer prospects than those whose career choices were essentially intellectual. They know that unless they rapidly move toward their goals, they will never make it big, much less keep their often very-well funded jobs, because usually they are backed by "soft" monies that suddenly might disappear, and if progress is not repeatedly proclaimed, the ax will fall to allow some new pressure group to have its chance. So the atmosphere aboard these massive national efforts can be most tempestuous, with conflicting approaches subject to great skepticism, if not scorn, by their respective opponents, and endless jockeying for next year's appropriations being a major aspect of daily life. Such programs thus become naturals for scientific messiahs, because with charisma at the top, the acquisition of next year's budget becomes a much more certain affair.

Now of course there are many problems where science may have a real chance to help mankind. Here you may ask whether a cause can ever be good enough to generate small worlds of selfless scientists where the common good is the overwhelming consideration and thoughts about priorities receive little attention. In fact, don't we already have such an example

in the recent "Conquest of Cancer" legislation? It not only provides to cancer research resources unimaginable several years ago, but also tries to
specify how dispassionate advisers can oversee its swift running. I'm
afraid, however, that here again we have an unrealistic pipe dream and
that insofar as we are dealing with the process of scientific discovery, the
sociology of cancer research will not show any striking differences from
other branches of science. Even if we disregard the tons of lousy cancer
research still being done and concentrate on the first-rate variety, which
for the first time is beginning to roll out in large masses, we still must deal
with the usual complicated mix of cooperation and competition.

Again, there does not exist any practical way to ensure that two talented people never want to solve the same problem. At a given moment,
certain areas will look more glamorous and about to crack, and I see no
practical way to push first-rate minds into less attractive approaches.
Moreover, some duplication of effort may be the only way to reach a goal
within a certain time span. No two minds ever take exactly the same path,
and placing all your marbles on one person's intuition never makes sense
if you have the money and want to move quickly.

Though the National Cancer Institute has many large-scale projects
set up specifically to bring people together so that they can pool their data,
not all labs want to operate that way. I even know of first-rate labs that
occasionally operate on the "need to know" procedure favored by government security agents anxious to keep key information from getting to
inappropriate offices both within and without our national government.
Scientists may only be let in on what their colleagues in the next room are
doing when such facts become necessary to let them proceed onto the next
crucial experiment. Such behavior, of course, is at times a consequence of
fears that someone will use your preliminary results to beat you to the
punch. But it may also aim to prevent the spread of hot facts that cannot
later be substantiated. No one may take you seriously if one day you
appear to say one thing, but the next week blurt out the opposite. Sticklers
for not making mistakes often compulsively resist letting out facts that
cannot be backed up with a written manuscript. Irrespective of its origin,
however, secretive science has the great limitation that you often do not
try out your ideas on your friends to see if they can spot a flaw that you
may have missed. So, unless you are very good, mistakes can go on longer
than necessary and the quality of your science goes down.

The question thus becomes posed whether obsessive secrecy is delaying the day when we can at last understand the molecular basis of cancer. If so, the public has every right to feel betrayed with its so far unwavering support for cancer research degenerating into the cynical attitude now directed toward almost anyone who makes a business of working to better the human condition. My guess, however, is that, at worst, these manifestations of the secretive urge will have only a marginal effect on the final result. On the one hand, secrecy obviously harms when it keeps people banging away on unproductive leads for longer than necessary. On the other hand, perhaps an equal number of scientists would lose their effectiveness if they did not have the peace of mind produced by limited secrecy. Unless you have unharried time to reach conclusions by yourself, unhampered by the thought that the next minute will bring a telephone call from someone doing the same thing, you can never acquire the self-confidence that comes from the realization that you can think through difficult problems or have the persistence to stick with an experiment until it is solved. Some forms of solitude are necessary for self-identification, and we would seriously misjudge the human temperament to suggest that we always work best in the company of others. Of course, in many instances, teams of two or three work well together, but they're often most effective if they collect different skills and members are not interchangeable.

A natural limitation thus exists on the numbers of people who can effectively cooperate with each other; once groups get bigger than a certain size, they tend to fragment. Two is the ideal number for most projects, because you always have a partner to keep you going when your morale temporarily lapses, while there is no opportunity to be the odd man out when your talents are no longer needed. Very difficult projects nonetheless may only be attackable if many more hands are available, but even so, large groups will stay together only as long as each member feels unique and not used by the others. Within the same walls, it never makes sense to have too many people with the same general objectives. As long as the expenses incurred in the duplication of facilities do not overwhelm us, it is instead better to set up a large number of different research groups working in different spots. This choice, of course, leads to the specter of duplication, but the advantages of feeling that you are master of your own fate will more than compensate for the tensions generated by the resulting competition.

Frequently, of course, you find that you can not only learn from, but also like, your competitor, and if you meet often enough at meetings and through visits to each other's labs, the close friendships that develop will lead, if not to jointly conceived experiments, most certainly to agreements to publish your data simultaneously. The fact that one of you may have reached a conclusion several weeks earlier matters much less than the fact that both of you had the sense to work out ways that would eventually culminate in the desired answer. Unfortunately, such gentlemanly agreements often pan out better when the objective is minor and not likely to attract many of your peers' attention. When you suspect that the correct answer might spectacularly alter current thought, the temptation is bound to arise to question the desirability of sharing rewards with someone with seemingly less credentials, particularly if he has a long history of jumping in on hot problems that only the hard work of others has made capable of quick solution. So, while many highly compassionate individuals exist and never behave in a dog-eat-dog fashion, the exact behavior we can expect from most colleagues is likely to depend upon the particular situation and, more often than not, passions will temporarily flare up and curse words will express the temper of the moment.

So I see no way that the rat-race aspects of much of high-powered science can vanish in the foreseeable future. To think otherwise would be to go back to the utopian ideals of the commune, a concept almost as old as modern man and whose periodic revival leads always to the rediscovery that selfishness often wins over altruism. But this realization does not mean we should be tempted to give up the scientific life. Initially, we chose it because the unknown in science was irresistible, and, more than likely, in other worlds we would be not only out of place but very bored. Moreover, other realistic career choices would not allow us to escape the competitive pressures of the modern world. No matter where you land, you have to do something well, and that will never be easy. Each of us must therefore somehow put our scientific dreams in line with the balance sheet of our limitations and capabilities, because if we do, we may perpetuate the excitement of our youths and still be able to anticipate each new day as a unique opportunity to follow our heart's desire.

Science and the American Scene[1]

1975

We come here today to renew our faith in the future; that of the Madeira School, that of American women as agents of intelligent action, and that of the American nation to base its existence upon the truth that comes from observations and experiments. In the building of this new center for the teaching of science, the Madeira School is making the statement that science is an indispensable aspect of the education of women as well as of men. Not only are the facts of science inherently important to our notions of what we are, and how we relate to the universe around us, but they have also provided the intellectual framework for the development of much of our modern civilization.

Equally important, the adoption of the scientific approach makes us realize that the way we live is the result of our own biological and cultural evolutions, not the outcome of the actions of supreme deities whose arbitrary whims could either give us great fortune or cast us into hopeless despair. If the world peoples become better fed, housed, and educated, it will be because of our strength and wisdom; conversely, if our Western-based civilization begins to lose its vigor, it will be our making, not that of malicious Jovian spite.

We thus have to respond to the situation that our nation no longer has its past self-confidence, and we wonder aloud whether we have lost our touch. Our industry seems unable to renew itself, and even the most American of all symbols, the automobile, is a dinosaur-like object, unable to move with the times. And we cannot comfort ourselves that we have

---

[1]This address was delivered at the Madeira School, a girls' school in Virginia, at which Watson's nieces were students.

merely fallen because of bad luck. It has not been the throw of the dice that today gives cities like Detroit such a sick spirit. Rather it has been the blind failure of the leaders of not only our Industry but also our nation to realize that the world's oil and gas supply is being rapidly depleted and that its price must dramatically rise. Though it is politically fashionable today to fault the Arab sheiks for daring to act like Texans, the primary blame lies with us as a nation in not realizing that our natural resources are limited and that the present outlook for the American people is inevitably downward if we do not change our course.

But this we have so far avoided, and the general belief still prevails in our land that provided we spend enough money, science and its resultant technology can save us from frugality. We go on hoping that the sloppy exuberance that characterizes today's "throw-away society" can continue into the indefinite future. But here we must realize that science, like any other human endeavor, has its limits, and there is no inherent reason to believe that even unlimited money for science can save us from the consequences of years after years of human folly. So the concept of economic growth as measured by consumption of material resources may no longer make sense, and as a nation we may have to strive for happiness in ways that do not automatically demand the still further depletion of our heritage from the earth.

In particular we must not go wasting more and more, opiated by the belief that our scientists and engineers can quickly come up with one or two all embracing solutions to our growing shortages of natural resources. Instead we must both simplify our desires as well as begin to place our bets for new sources of energy and raw materials in many directions. Thus while the breeder reactor might some day be able to be developed to produce unlimited amounts of nuclear fuel, it is not clear that the consequential massive development of nuclear energy could ever be separated from unbearable exposure to nuclear blackmail. And while we all hope for wide-scale utilization of solar energy, to follow the enlightened example we are here to dedicate, we would be most silly to say that it can be the whole answer and that willy nilly we must try to block the development of all new types of oil-, coal-, or nuclear-based power plants no matter where they may be sited.

This inability of science and technology to do what we want on a neat time schedule is of course not limited to the energy problem. We must not naively think, for example, that we can induce more and more lung cancer by smoking and asbestos, and then by spending more money for cancer research come up with its cure. This does not mean that we should not try very hard to do the research that may lead to cures for this horrific disease. But we do everyone a great disservice, especially our smokers, if we imply that what is lacking is money. Such problems may be too hard to crack within the short term, and only when much much more of the seemingly pure and so-called irrelevant knowledge has been accumulated can the answers emerge.

Now for any truly educated person, what I have just said is common sense and you may ask why here I emphasize the limitations of science, not its endless potential for help to mankind. My answer is that I sense an increasing unbalance in the relationship between science and the American people. Because we have done so much, it seems natural that we should do even more. And I'm sure that in the long term, we shall produce further wonders that shall dazzle even our most advanced cynics. What I worry about now is the framework of time. We all too often talk of solutions by the next year to decade, while instead we should be framing many of our main dreams in terms of next century or two. Though we went to the moon in this century, it may only be in the next that we can come to grips with most cancer or have a chance to do anything with schizophrenia.

Yet, for example, an increasingly real part of the National Science Foundation's (NSF) budget goes toward programs designed to relieve short-term human needs. And not surprisingly, most such applied efforts are badly floundering because the NSF has taken on more than it is today capable of. Yet this is not really the fault of the NSF, most of whose leaders never wanted these so-called "relevant" programs to start with. Instead the blame lies with us as a Nation. Because the American public as a whole still believes that a knowledge of science is best left to scientists and most certainly not to young ladies, we remain as a people dismally ignorant of science and its mode of operation. And so countless decisions, in which the utilization of scientific facts is a key ingredient, are made by persons totally ignorant of its vocabulary, suspicious of its better practitioners, and loath to take advice for fear that they will be asked to run counter to short-term political common sense.

Thus, the question that must be faced is, Will we as a nation always have to live with the fact that many, if not most, of the major long-term decisions that involve science will continue to be made by people unqualified for such responsibilities? In our beginnings, when we were a simple agricultural nation, whose culture was largely limited to the Bible, we could be responsibly ruled by men of God, the law, and the leger. But now that we are so much a land of science and technology, it makes no sense to have us led as if these subjects were but minor ingredients in our national mix. Instead the politicians who make the decisions must be truly educated and so conversant with science. Here we must be clear that we should not count on their scientific surrogates to save the situation. We shall not effectively come to grips with science by merely inserting scientific advisors into the executive and legislative branches. As advisors, they will always be considered guests, and we all know too well the fate of guests who do not agree with their hosts.

Thus, our true salvation may only come when we begin to regard science as an indispensable part of the education of all our citizens, not topics limited only to those compulsively bright kids whose enthusiasm for perpetual learning makes them congenetically unable to put up with the polite banters that characterize the successful lawyer or banker or businessman. But as long as our leading secondary schools regard the lack of a special aptitude as a valid reason for the avoidance of science, we will continue to find that the majority of our better secondary school students will go on to the universities totally ignorant of what science may be capable of and firmly accepting the fact that in their lives science is a lost cause that best can be forgotten by the study of torts or the midnight reading of the late Victorian novels.

So the time has more than come when our survival as a great people depends upon the extent that science is a key ingredient in the training of all our better students. So we should now be very happy that the Madeira School begins to take science most seriously. The strongest statement that Madeira can make is that women have a real role to play in preservation, if not advancement, of our civilization.

The Necessity for Some Academic Aloofness

1979

The stereotype of the pure eccentric professor, indifferent to the world about him and motivated only toward the intellectual advancement of his own field of research, is convenient to perpetuate. For the academic it provides the perfect way out from the more than occasional necessity of appearing concerned with the many pressing problems of the imperfect societies in which we all live. A monomania of some form or other is a prerequisite for most intellectual achievements, and trying to do too much all too often is a recipe for future lack of distinction. And even if we are no longer passionate about our intellectual careers, we tend to be reluctant to announce publicly that we are ready to help our neighbors. This would imply that we frequently have ideas that have social consequences. We do not need to be clever to realize that the appearance of goodness is far from correlated with the actual accomplishment of good works. Far better to maintain the illusion that we always have esoteric ideas in mind.

This way of proceeding is also not without its advantages to our politicians, the ultimate source of almost all our academic money. They need to be seen as practical leaders and do not want to be unfairly tagged as unworldly by too much association with minds who cannot be counted on for reliably consistent answers. As the leaders of our countries, they have to devote almost all of their waking moments to today's decisions, leaving to their dreams the long-term future. They know as administrators that the problems of the immediate moment cannot be pushed aside in favor of schemes that perhaps never should have been generated. Most untested new ideas never go anywhere, and old remedies, imperfect as they may be, have that marvelous quality of being predictable. Inherently, none of us want to live with too much uncertainty, yet we as university types seem often to push for new approaches merely because we are bored with the past.

Most new ideas of intellectuals have to be seen for what they are, namely, academic exercises usually put forward without the thought that they might be taken seriously by the outside world. Without the fear of visible failure, the mind meanders faster and further, sometimes toward more practical directions than initially perceived. This is the way it should be. If we believe that our futures as viable societies demand infusions of the really new, it makes sense for the most part to separate the generation of new ideas from the job slots where the consequences of failure are too clearly seen. This is why universities and institutes of pure research must exist as partial oases from the real world. Only in somewhat unhurried atmospheres will we take the time to assimilate the key features of other cultures or to scrutinize carefully the tenets of the moment for their viability over the foreseeable future.

Thus, there will, of necessity, always be a sense of quiet unease in an academic-like environment. The desire of the students and younger faculty to reexamine, if not overturn, the inadequate ideas of the past will often be in conflict with the purposes of the more senior professors who themselves a generation before may have come to power through their ability to think differently. A major factor determining the quality of a given institution is thus the ability of its faculty to reward intellectual success even when it leads to the effective academic redundancy of many of its older members.

Happily, the tradition of rewarding intellectual novelty now dominates major universities in every society—that is, as long as the ideas promoted are not thought to affect the political (social) stability of the society concerned. Here we should not delude ourselves that these exceptions are all that rare. No one, for example, can imagine a promoter of Marxist economics becoming a professor in today's Chile. And even at Harvard, Marxist economists never seem to make it.

For the most part, however, natural science has been allowed to evolve on its own terms. No society that wishes to advance itself in the modern world can reject scientific progress, and we now find, virtually independent of the political system, that the content of science is left to its practitioners and not imposed from above. There is, for example, no capitalistic form of molecular biology. No matter in which country we look, molecular biologists are apt to have the same common objectives. To be sure, there have been serious efforts to turn the clock back: Witness the debasements of Russian biology when Stalin decided that Mendelian genetics posed a threat to communist doctrine on why different individuals have different potentials for future success; or the much less successful attempt by the Nazis to label relativity research as Jewish science; and the rear-guard action by fundamentalist American religious groups to prevent the teaching of Darwin's evolutionary theories.

All in all, however, the more meaningful fact is the paucity of such restrictive political actions during the development of modern science. The efforts of the Catholic church to muffle Galileo in the 16th century fortunately have not had too many subsequent parallels. Since then, the growing clear correlation between the existence of a vigorous free science and the generation of technological advances has led more and more nations to organize their support of science in ways that maximize their scientists' control over their own destiny.

Here we cannot overemphasize too much the necessity for the promotion of pure science. Through it are generated the unexpected new facts that later can lead to applied developments such as high-speed computers or the drugs that control our fertility. The outcome of untried scientific experiments are remarkably unpredictable, and we are constantly amazed by how different nature is from what our experts of the past have prophesied. And despite more prognostication that the end of new knowledge is in sight, there is general agreement that there has been no real slowing down in the rate of which important new scientific concepts are established. Most, of course, have no immediate practical consequence. But when one does, say in the development of semiconductors by the solid-state physicists, the payoff may be large indeed.

While for a short time a growing society may try to live off the new facts produced by others, this approach appears not to be a recipe for long-term worldwide leadership. Witness that the United States only became the major world power when it was the undisputed leader of world science; also note that the long-held conception in the United States and Europe—that the Japanese get ahead by copying the West—is almost certainly false, reflecting more our inability to speak their language than any critical evaluation of the way they educate their students and encourage the development of new ways of thinking.

What almost seems inescapable is that the deeper we understand the ways of nature, the more capable we are to use it for human benefit. Over and over in retrospect we see that we never had a realistic chance to solve an applied problem until our scientific base was raised. Only, for example, by discovering that there were three immunologically distinct forms of polio virus could we move beyond the inadequate iron lungs and truly conquer polio.

We must also realize that pure scientists, almost without exception, have no reluctance in exploiting their observatons for practical use. The fact that most of us do not get involved in practical matters is not a matter of principle, but usually reflects the perception, sometimes mistaken, that we do not have anything unique to offer the outside world. But when we feel that a new observation has much wider application than a place in a scientific journal, we are unlikely long to ignore the fact that we might enhance our careers, if not our practical fortunes, by moving our thoughts to the outside world. To show this, I can best draw on my experience in molecular biology, a field that until recently was thought to be one of the more esoteric of the major sciences. Two major examples now stand out.

The Genetic Engineering of Microorganisms to Produce Medically Useful Hormones and Drugs

The now over 30 years of intensive research on DNA, the chemical compound that carries the genetic information of our genes, and on the fundamental genetics of bacteria, has led recently to the development of new procedures in which bacteria can be used as effective factories to produce large amounts of human proteins that now are difficult to prepare in medically satisfactory amounts (e.g., the growth hormone and, hopefully soon, the infinitely more illusive antiviral compound interferon). While vague proposals for genetic engineering have been talked about ever since DNA was discovered to be the gene in 1944, this field became technologically possible only following the unexpected isolation in the early 1960s of a new class of enzymes (the restriction enzymes). These catalysts cut DNA molecules at specific sites and thus allow the routine test-tube generation of "recombinant DNA" molecules, say partially of human origin and partially of bacterial origin. As first shown in 1973 by Herbert Boyer and Stanley Cohen, these hybrid molecules can then be put into receptive bacteria where they multiply as functional minichromosomes and lead to the production of not only proteins of bacterial origin, but also those corresponding to the specific human genes from which they have been constructed. Already three human hormones (insulin, somatostatin, and pituitary growth) have been synthesized by these procedures, and over the next several years we should witness an explosion of such examples, many with great commercial significance. This is not the place, however, to give the technical details of these procedures. Much more interesting to this gathering are the facts that:

1. There is no way these procedures could have been worked out by applied scientists without prior intense development of pure molecular genetics and bacterial genetics.
2. The first successful recombinant DNA experiments were carried out at two major academic sites (University of California Medical School at San Francisco and Stanford University Medical School) by scientists whose careers until then had been in pure science.
3. The first use of these procedures was to further our knowledge of pure science, and already several major advances in understanding the structure and function of the genes of higher organisms have occurred that would have been impossible without recombinant DNA.
4. The initial developers of the recombinant DNA methodologies realized from the start that there could be major commercial applications, and in due course they became consultants and stockholders in several of the new companies that were formed to exploit the technology. They also saw to it that patents were taken out by their respective universities to protect their rights, as well as their universities, in the development.
5. No major pharmaceutical company quickly moved into this work, in part because on the whole they had seen no previous need to emphasize applied developments in either bacteria genetics or nucleic acid enzymology. So they didn't possess strong internal lobbies to push the

advantages of recombinant DNA. They were also afraid to be labeled as genetic engineers, knowing that the public badly confused work with bacteria with actual experimentation on humans. However, with the now much diminished public concern about recombinant DNA, we may expect the pharmaceutical giants to try to catch up, often by buying some of the newly formed recombinant DNA companies.

Growing Industrial Exploitation of Monoclonal Antibodies

Until several years ago, antibodies either for vaccine production or for medical testing or scientific research were made by injecting the appropriate antigens (e.g., diphtheria toxin) into suitable laboratory animals (usually rabbits and guinea pigs and when large vaccine amounts were needed, sheep and horses). All such resulting antisera are heterogeneous, containing antibodies of varying degrees of strength directed against different chemical groups on the surface of the respective antigens. Long desired, but not apparently easy, was any way to make preparations of identical strong antibodies all directed against a single antigenic group. In the 1960s this goal began to look increasingly distant as we began to learn some of the basic facts about the cellular basis of the immunological response. In particular, the immunological response of a single animal usually involves the selective synthesis of large numbers of different lymphocytes, each of which makes its own unique type of antibody. Hence the heterogeneous nature of all available antisera.

It thus became obvious that the only way to produce single specificity antibodies was to somehow find the conditions where single antibody-producing cells could grow in culture like bacteria. By starting such cultures with single lymphocytes, all the resulting antibodies would be made by genetically identical lymphocytes and therefore possess identical antigen-combining sites. All such efforts, however, invariably failed to tell us that the normal antibody-producing cells lack the ability to multiply indefinitely. This is a great contrast to the behavior of the abnormal antibody-producing cells that generate the cancers we call myelomas. Such cells are effectively immortal and multiply without stop in culture, a condition no doubt related to their cancerous state. Their existence, however, gave to George Kohler and Cesar Milstein, working in Cambridge, England, the idea that if they fused mouse myeloma cells with normal anti-body-producing cells from the spleen of a mouse that had been immunized against sheep red blood cells, the resulting hybrid cells might continue to produce the desired antibody against sheep red cells, as well as to become immortal. Five years ago, the first successful such experiments were done, giving rise to clones of mouse cells that provide antibodies of desired single specificity (monoclonal antibodies). Since then, monoclonal antibodies have been made against large numbers of different antigens, and this procedure, because it leads to highly reproducible antibodies of high specificity, may soon supplement all current commercial ways of making antibodies. Here it is relevant to emphasize the following:

1. This technique could not have developed without the past two decades of solid advances in understanding the cellular basis of the immune response. Equally necessary was the finding (in the mid 1960s by the English cell biologists, Harris and Watkins) of routine ways for fusing cells.
2. All the key steps were done by pure scientists, this time working in a laboratory for pure molecular biology, supported by the British Medical Research Council.
3. The most important early uses have been in the area of pure immuno-biology and cell biology, giving to this field new armadas of specific antibodies directed against biologically important molecules on cell surfaces.
4. The idea seemed so simple that no thought was given to patent protection until it was too late. Nonetheless, the extraordinary power of the monoclonal antibody procedure has quickly led to multiple commercial uses. As with recombinant DNA, several companies have been formed to quickly exploit it. The fact, however, that most major pharmaceutical houses have made in-house immunological strength has quickly led to their moving to improve their vaccine (medical technology) capabilities through the monoclonal procedures.

Both of these examples are textbook examples of an advancement in pure science leading inescapably to the public good through the production of products not previously effectively available. What is novel here is that molecular biology, prior to these discoveries, was regarded as one of the purest of the sciences. Until a year or so ago, almost none of our students ever thought about jobs in industry nor were my peers boosting their incomes by being asked by industry for their advice. Now our lives are very different, and many of the peers that I have most regard for as scientists are shuttling back and forth from their industrial backers. And, of course, they are no longer saying that the doers of applied sciences are no better than mercenaries for the rich.

I, therefore, conclude that the main factor that keeps academic thinkers aloof from the real world is not a matter of principle, but the lack of common interest, with most encounters likely to make both sides somewhat edgy. While we as intellectuals have ideas as our chief currency, those in industry, business, or agriculture must focus at the economic production of the worldly goods without which our societies would quickly crumble. Our short-term roles are thus so different that very little good is likely to come through too frequent intermixing.

On the other hand, when we do have relevant answers, then we are not the impractical souls of common perception and can make ourselves effectively heard. Our physicists, for example, have never hesitated from entering into inherent political battles about nuclear weapons and how they may be controlled. They know they will rise or fall along with the rest, and their appearance of absent-mindedness must necessarily be dropped when they are needed.

At the same time we must never forget what we can do best. In our searching for ways to understand better the world around us, we are a vital, if not *the* vital, key for ensuring that the various civilizations of men can long prevail over the inherent chaos of the physical world about us. Thus, we have as much to lose from false humility as from elitist arrogance. In working toward the reconstruction of our societies, we must never forget that ideas are not only beautiful, but necessary.

Striving for Excellence

Let me start with the difference between excellence and excellent. "Excellence" is clearly something to strive for. But excellent doesn't go far with me because I am a product of the Chicago public school where we were graded with F, G, E, and S—fair, good, excellent, and superior. If you were only excellent, you were not on top. I find that when I write recommendations for former staff members or students and if I say someone is an excellent scientist, I mean he probably can hold the job I am recommending him for. So it depends on where you are sending the letter. But it doesn't say that much. If you say he is unusual or out of the ordinary, that may frighten people off. Most certainly you can't say that he is a genius, because high schools are filled with geniuses and you never hear of them afterward. If you really want to say that he is top rate, then you compare him with someone whom everyone knows is super.

Now back to excellence. For me as a writer it means trying to come up with an idea or book that people I respect will want to read or talk about. The question then becomes of learning how to produce novel thoughts or words. Here I believe that imitation is very important. Even though imitation might initially sound the cheap way to go, it is only a question of whom you imitate. If you are trying to mimic someone very good, you really can't pull it off, but what comes out might nevertheless be fairly interesting. But all too many people are taught to strike out on their own before they know how the best of the past was achieved. When I was 23, I went through a phase where I wanted to understand how Linus Pauling thought well enough so I could write a paper in his style. And several years later I read *The Great Gatsby* and I began to dream that I might produce a novel with the class of Fitzgerald. Being imitative should not stop with your youth. For example, when at Caltech I watched the way George Beadle ran its Biology Division; later, when at 40 I became the Director of the Cold Spring Harbor Lab, I tried to measure my performance by his past high selfless standards.

Equally important, when growing up you should not be focused on trying to get the respect of your peers. Then, by definition, they are too young to know what is up. Instead you must always look to older people. Of course, when you finally grow up and are past 50 you can't look up to senility. Fortunately, if you try hard there are still some very clever people in their 60s or 70s. I am just beginning to face that problem. But when you are young, say between 15 and 20, how do you get the courage to rise above your environment to think differently? Clearly I had a number of breaks when I was very young. For example, I had no respect for my grammar school class. It started with the election of 1936. There were 40 of us in the class, and 38 had Alf Landon sunflowers pinned to them.

So I grew up knowing that there was no point trying to behave like others. Not that our middle-class neighborhood was rich. It wasn't affluent at all, but still all those little sunflowers. Then my father read *The New Republic* and the *Nation,* having just stopped bringing home the *New Masses* and things like that even though he had come from an Episcopalian family. As a young man, he found the worship of God antithetical to reason and took joy in reading books that proclaimed religion the enemy of the rational mind. So with the help of my parents, I realized that authority was not necessarily to be respected. Also, the boss of the Chicago schools was its superintendent, Mr. Willard Johnson, who wrote virtually every textbook we had to read.

When you reject authority, you first have to have someone point out to you the real stinkers. Being against everything is a sure way to get nowhere. When I see someone so basically negative, I block him out from my mind. You have to be very selective in finding your models. I was lucky because I had my early formative years when Franklin Roosevelt was president. He gave marvelous speeches. I will never forget his use of the phrase "Martin, Barton, and Fish." With those three words the Republican plutocrats were never going to elect Wendell Willkie. Roosevelt really believed in the English language and used super speech writers like Robert Sherwood and Sam Rosenmann. So, even though there was not much style on the southside of Chicago, I could hear it on the radio. Almost as important to me was Churchill, like Roosevelt a born Tory. But they both respected intelligence. Your could just see it in their use of words. And then there was a war to be fought. Life then was fairly simple. The Depression and War. But then I had heroes to help me.

It also helped that I had an uncle who was a physicist. A modestly good physicist who became a professor at Yale, so maybe academic success was in the family. And another uncle, Dudley Crafts Watson, the lecturer on arts at the Art Institute, who often wore a smock and was a whiz in drawing local PTA audiences. More importantly, he helped raise Orson Welles, who had done something more than unique with his *Invasion from Mars* and did it when he was very young. Uncle Dudley had a monocle, but that was invisible when he gave his radio talks on WGN.

Thus, there was a tiny trace of bizarreness in my background, and because I had no prospect of being successful in a normal way, beginning in my early teens I took delight in the charmingly eccentric and loved the movie *Pygmalion*, which I could not see often enough. Besides listening on the radio to Roosevelt and Churchill, the main formative moments in my early life came from films. I went to the movies all the time because they showed you something outside the southside of Chicago. Then we lived about two miles from the Carnegie mills. Luckily our house was filled with many hundreds of books including *USA* by Dos Passos and *Studs Lonnigan* by James Farrell. Dad's books told me more about America than I would learn from my daily life.

Starting in college, I tried as much as possible to be among really bright people. Soon I found that they generally had as little idea as I did about how to solve the key problems ahead. But it makes sense to agree with your professor when you first enter his lab. You must not then slavishly follow his directions too long because he is probably equally at a loss as to what to do next. Soon you will have to strike out on your own.

After the double helix was found, I went off on my own to Caltech and was unproductive trying to guess the structure of RNA when it was unguessable. I knew that many outsiders would now think that Francis Crick was the main DNA brain, because he was clearly superbright while I was a dreamer who often got others to think for me. After two such years I went on to Harvard, largely because it was coeducational and I did not want two reasons for being unhappy. There my reputation was slowly restored by its bright graduate students. I almost never put my name on their papers or made them think they worked for me. So they worked even harder. It also helped that my theoretical physicist friend, Wally Gilbert, who I first met when we were both in England, joined me as we were finding messenger RNA. Wally was, and still is, extraordinarily intelligent as well as hardworking. Our 1960s lab, together with the ones Francis ran with Sydney Brenner and Monod ran with Jacob in Paris, were then the best shows going in what had by then become known as molecular biology.

After messenger RNA was a solid fact, I moved orthogonally away from molecular biology rat racing to become, I hoped, a serious writer of books. Initially I planned an elegant little successor to Schrodinger's *What is Life?* that I would call *This is Life*. With time it grew into a medium-sized, almost-text, *The Molecular Biology of the Gene*, that eventually sold enough copies to make me almost indifferent to my underpayment as a Harvard professor out of phase with its president. In writing it, I gave it an intellectual flavor that made it more than a meticulous rendering of the new facts of the past decade. After it came out many friends wrote me that they liked it very much. I was very relieved by having again done something of excellence.

At the same time I was writing *The Double Helix*. In the spring of 1962 I had given a public lecture in New York on how the DNA structure was actually solved. It provoked much laughter and I knew I had to put it in writing. Initially, I daydreamed that *The New Yorker* might print it under the rubric "Annals of Crime," because there were those who thought Francis and I had no right to think about other people's data and had in fact stolen the double helix from Maurice Wilkins and Rosalind Franklin. Some months later, I met on Paul Doty's porch *The New Yorker* writer Daniel Lang, to whom I excitedly revealed my plans. He wasn't interested, but I saw no point in taking his advice because I had by then learned that Radcliffe girls opted for writers over scientists. I had the first chapter typed by a very fey girl with incredible cat-like blue eyes. My opening sentence, "I have never seen Francis Crick in a modest mood," did not, however, affect her as it did me. With it I knew I had a book that must be finished. But I knew I best not finish it too soon because we hadn't received the Nobel Prize and, if it was published prematurely, we would never get it.

My objective from the start was to produce a book as good as *The Great Gatsby*. I was then also much impressed by Christopher Isherwood, especially his *Mr. Norris Changes Trains*, a now largely forgotten, marvelously perceptive description of slightly seedy people. I had also read everything by Graham Greene, *The End of the Affair* and *The Heart of the Matter* strongly affecting me. I also had a good story to tell, and thought that with effort it might read like a Fitzgerald novel. Jay Gatsby was partly a fraud and while I hadn't a bootlegger past, there were some scientists who thought I wasn't much above it. And so, I wanted to put the whole story down with the ambiguity about motives in clear focus.

Very important was stumbling early upon the title *Honest Jim*. Kingsley Amis' *Lucky Jim* had made me laugh uncontrollably, and then there was *Lord Jim*. So maybe I might be able to fashion a book that would some day be placed in some class as the two other famous "Jims." To do this I had to make my friends worth knowing about in a realistic fashion. They had interested me, so I thought maybe they would interest others.

In America you are seldom taught enough about word usage and my English years were essential in my becoming an accomplished writer. A few people here do occasionally use words cleverly, but it is not a national virtue. We are told that if we are sincere, it doesn't matter that we limit ourselves to words that we have known since, say, the sixth grade. After I had a chapter tightly finished I would give it to one or two good Radcliffe friends who were bright as well as good-looking. If they laughed, I would feel good and go on happily to the next chapter.

So, the writing of *The Double Helix* was largely a pleasant chore but Francis at first did not want it published, saying that it was much too light a work to convey accurately how the double helix was found, and most certainly it should not carry the label of Harvard University Press. With these sentiments Harvard's President Mr. Pusey agreed, and so lots of money that the University would have made went instead to The Athenaeum Press.

I was pleased, though surprised, that it became a best seller. I did not anticipate, however, that so many readers would treat it as serious literature, and I was satisfied especially with the comment of one reviewer that I could get a second Nobel Prize if I wrote a couple of more books of similar quality. So, I have plans for and already have written the opening chapters to a successor of *The Double Helix*, which is to be called *Calculated Madness*.

Succeeding in Science: Some Rules of Thumb[1,2]

$\diamondsuit$

1993

"To succeed in science, it's not enough to be smart—lots of people are very bright and get nowhere in life."

REFLECTIONS

To have success in science, you need some luck. Without it, I would never have become interested in genetics. I was 17, almost 3 years into college, and after a summer in the North Woods, I came back to the University of Chicago and spotted the tiny book by the theoretical physicist Erwin Schrödinger. In that little gem, Schrödinger said the essence of life was the gene. Up until then, I was interested in birds. But then I thought, well, if the gene is the essence of life, I want to know more about it. And that was fateful because, otherwise, I would have spent my life studying birds and no one would have heard of me.

Instead, I became absorbed with one of the defining questions of the twentieth century: What was the gene? And then I got a second break: I was turned down for graduate school by Caltech. Why would they be interested in someone whose principal college work was in birds? So I went to Indiana instead, and my professor was none other than Salvador

---

[1]Adapted from an after-dinner talk at the Cold Spring Harbor meeting to celebrate the 40th anniversary of the double helix.

[2]At least two distinguished biologists have written books of advice to the aspiring young. The first appeared late in the nineteenth century and was written by the father of neurobiology, Santiago Ramón y Cajal; the second, actually entitled *Advice to a Young Scientist*, is by the British immunologist and Nobel Laureate Sir Peter Medawar.

Luria. His belief in me gave my early interest in genetics a big boost. And unlike Pasadena, Bloomington provided girls and basketball.

But to succeed in science, you need a lot more than luck. And it's not enough to be smart—lots of people are very bright and get nowhere in life. In my view, you have to combine intelligence with a willingness not to follow conventions when they block your path forward. For me, that meant giving up Luria's way of doing things, even before I had finished my Ph.D., and searching for my own way. And it meant doing lots of other things a little differently than most people. And these have become my rules for success.

Learn from the Winners

Take the first rule: To succeed in science, you have to avoid dumb people (here I was still following Luria's example). Now, that might sound inexcusably flip, but the fact is that you must always turn to people who are brighter than yourself. It's like playing any game—tag or tennis. Even as a child, I never liked to play tag with anyone who was as bad as I was. If you win, it gives you no pleasure. And in the game of science—or life—the highest goal isn't simply to win, it's to win at something really difficult. Put another way, it's to go somewhere beyond your ability and come out on top.

Take Risks

Which brings me to my second rule: To make a huge success, a scientist has to be prepared to get into deep trouble. Sometime or another, people will tell you that you're not ready to do something. Take my early career. The zoologist Paul Weiss, from whom I had learned about invertebrates, had a good brain but he was lacking in vision. That didn't help me, because he was in charge of my fellowship in Europe. He took away my stipend when I decided to move from Copenhagen to Cambridge. In his mind, I was not prepared for crystallography as a biologist. He was right, of course. But the only way I was going to make the next breakthrough in genetics was through X-ray diffraction analysis, even though most of its practitioners thought DNA to be an unrealistic goal. If you are going to make a big jump in science, you will very likely be unqualified to succeed, by definition. The truth, however, won't save you from criticism. Your very willingness to take on a very big goal will offend some people who will think that you are too big for your britches and crazy to boot.

Now this act of ignoring the assessments of those who seem to have the power to control your fate can be traumatic. Often it entails rejecting your mentor, your lab head, or your department chairman. But to get where you want to go, you even have to be prepared to give up your second parents. You've probably already given up your real parents—that was a hard one—and now you have to give up your scientific heroes. This can be more than just personally upsetting. You can, if you're not careful, develop deep anxieties.

Have a Fallback

And that brings me to my third rule: Be sure you always have someone up your sleeve who will save you when you find yourself in deep shit.

Francis Crick and I were both in trouble at various times in our careers, but that never really stopped us because we always found someone who would save us. In Cambridge, both Max Perutz and John Kendrew stood behind us. John, for example, promised to let me live for free in his Tennis Court Road house after Paul Weiss had cut off my stipend. But I think of many bright scientists (our DNA-hunting competitor, Rosalind Franklin, was one) who, when they found themselves in trouble, had no one who would or could save them.

Luria first saved me when, as a mere graduate student, I outraged a prominent professor, Ralph Cleland, who wanted to make me take histology. I considered that course a waste of my time. And I said just that with Cleland in the room as part of my thesis committee. Luria was, of course, very upset with my bluntness. But he didn't join forces with Cleland and force me to waste my time doing silly drawings when I could be having fun with phages. So I went out into the world ignorant of histology.

Have Fun and Stay Connected

Which brings up my fourth rule: Never do anything that bores you. My experience in science is that someone is always telling you to do things that leave you flat. Bad idea. I'm not good enough to do well something I dislike. In fact, I find it hard enough to do well something that I like. And that brings up another reason for having people around who care about you—

you have to have people you can go to for intellectual help. Francis frequently went for assistance to his more-than-bright philosopher/mathematician friend, George Kreisel, who was noted for not respecting conventions. He seemed to have no close friends except Francis. But Francis frequently turned to him when the mathematics got difficult.

Constantly exposing your ideas to informed criticism is very important, and I would venture to say that one reason both of our chief competitors failed to reach the double helix before us was that each was effectively very isolated. Rosalind Franklin found small talk awkward and, until it was too late, did not realize how much good advice Francis would willingly have given her. Had she started to talk to him, Francis would have led her to use her facts to find the base pairs. And then there's Linus Pauling. Linus' fame had got him into a position where everyone was afraid to disagree with him. The only person he could freely talk to was his wife, who reinforced his ego, which isn't what you need in life.

While it is tempting as a young person to dream about going into science because you think ideas will let you escape dealing with people, once you are a scientist you must change your attitude. When you are in high school, it can be very comfortable to say to yourself: Why should I be with those kids that are commonly awful? But here's the truth: It's very hard to succeed in science if you don't want to be with other scientists—you have to go to key meetings where you may spot key facts that would have escaped you. And you have to chat with your competitors, even if you find them objectionable. I did that a lot. I knew almost everyone I needed to know, no matter whether they exuded goodness or badness. And it paid off.

So my final rule is: If you can't stand to be with your real peers, get out of science.

War on Cancer

The Academic Community and Cancer Research

1974

hat MIT is now doing in the setting up of this magnificent new Cancer Center is very sensible and wise. It is quickly responding to the facts that more cancer money is now available and that intelligent private foundations exist to create new research laboratories. You might say that any intelligent private university would do the same. But I fear that MIT's course is unique. It is the only nonclinical American institution that has responded to the "War on Cancer" in this grand way. But MIT must pay the penalty for acting more responsible. It must carry a greater burden in seeing to it that fundamental cancer research is carried out in a sensible way over the next decades.

To understand why MIT stands in this almost lonely position, we should look backward some 30 years to the origin of our current large-scale commitment aimed at the eventual banishment of cancer. Starting in the 1940s with the Committee on Growth of the National Research Council, there was the general consensus that if you were going to seriously study cancer, you had to generate more pure biology. That is, the science of biology was then not in a state where you could think sensibly about the cancer problem. So both the Federal government (through its various agencies) and the American Cancer Society realized that they had to do something for pure, irrelevant science· The Cancer Society, for example, funded my first summer (1948) at Cold Spring Harbor studying bacterial viruses (phages) under Salvador Luria. It was all too clear that biology must be nourished out of its predilection for descriptive analysis. That this might be accomplished if you learned the key messages of the new physics and chemistry that had come out of university science departments in the

1920s and 1930s and started applying them to the major problems of experimental biology.

Happily this revolution did occur, and between 1945 and 1965 biology was transformed from a largely descriptive nature into a science with ample analytical powers. At the same time the health lobbyists, whose enthusiasm had generated the initial flow of governmental and foundation cash, more and more began to think in terms of the human payoffs that soon should blossom forth. Here we must realize that 20 years (1945–1965) in the average person's life is a long time. Moreover, it is only when you reach 45 or so that you begin to have the ability to influence (lobby) the so-called older establishment. So after 20 years of influence you may be near 65 and realize that only a few more years may remain for you to see something happen out of your efforts. By 1970 we had reached that stage in the lives of our key cancer lobbyists, and because it was universally agreed that biology had come of age, it seemed almost natural, if not God-ordained to them, that the new biology should bring a halt to the horrors of cancer.

At the same time, those of us who had participated in the birth of the new biology—largely by doing experiments on the most simple cells, the bacteria and their viruses, the phages—began to realize that now was the time to move onto the world of the much more complicated cells of higher animals. With luck we might begin to understand the essence of embryological development as well as possibly coming to grips with the essence of cancer cells. But we worried that we might not have enough money, because no matter how we might proceed, the need for much more money, if not big money, seemed unavoidable. And so as the 1970s broke, there existed two independent pushes for lots of new cancer money. To be sure, they had two very different immediate goals, one to cure cancer, the other to understand it. In any case, our country was thought to be coming out of its greatest national debacle and both the clinicians and the pure scientists thought there should be money saved from the Vietnam War that could be used for cancer research.

For my part, I eagerly went down to Washington in the summer of 1970 to testify before the Yarborough Committee that the further study of the so-called cancer viruses was likely to be the most direct way to zoom in on the genetic basis of cancer. Though my advice was obviously parochial, because I myself hopefully would use some of the new cancer money, I felt sure that such money, if properly spent, could not fail to pay off eventually by telling us exactly the enzymatic basis by which a cell becomes cancerous. At the same time other witnesses, many more in number, testified that at long last we are beginning to cure by chemicals some forms of disseminated cancer (e.g., chorocarcinoma) that we couldn't cure before. Give us the money to set up additional large cancer centers like the Sloan-Kettering Memorial complex, and over the next decade we can cure even more cancers either by better use of anticancer drugs or through new immunological procedures. Though both types of testimony helped ensure the eventual passage of the conquest of cancer legislation, there is no doubt that it was the latter testimony about possible cures over the next decade ("a billion a year for 10 years will do the job") that created the climate for its unanimous passage and immediate funding. It is the lobby which says we may have real results in 5–10 years that carries the day. Talk about longer intervals and everyone gets bored—at least those over 50, which includes almost all of our Congressmen.

In fact, I used to ask for money for our Cold Spring Harbor Laboratory on that schedule. Give us the money to get further inside the animal cell and in 5–10 years we will begin to understand and may start to cure more cancer. But now after 7 years I find it hard to throw out the same old message, because the more we learn about normal higher cells and their cancerous equivalents, the more staggering the task I realize we have cut out for ourselves. We may easily have 20–50 years ahead of us as pure scientists before we can precisely say, for example, why a cell has become leukemic. This does not mean that nothing of consequence will happen sooner. In fact, the science of cancer goes increasingly well, and every new year, if not month, brings forth something we now momentarily think to be a fundamental lead. But we must be very careful not to confuse our day-by-day intellectual thrills with a thorough understanding of the nature of cancer. To do so is cheating, and while the rewards for perpetual enthusiastic optimism may be piles of soft money, the long-term result has to be loss of intellectual integrity. Nonetheless, when I say my long-term goal merely is only to understand cancer, not to cure it, I fear not only that I will be deemed socially irresponsible, but that I will find it increasingly hard to get money for a goal that does not directly aim to help the immediate sick or those who soon will be so.

Hopefully, of course, such fears have no basis because I don't know any responsible person who doesn't realize how much more fundamental cancer research may be necessary. This fact, however, by itself does not mean that as a nation we shall necessarily use the new cancer money in the best of all possible ways. To start we must realize that our plan of war, "The National Cancer Plan," is largely a public relations effort. Initially conceived as the flow chart for a Polaris Missile-type approach to cancer, it grew out of a number of meetings of cancer experts at Airlie House, a conference center in Virginia. Though their talks produced evidence of large numbers of possible leads into cancer, I find the final massively heavy document a total sham that carefully took care not to distinguish between the respective merits of the so-called "hot leads." Nowhere can you find directions to spend more here and less there. So at best it is a catalog of wishful thinking to which the director of the NCI can point in front of Congress and say "Yes, we know where we are going." But during my two years' experience on the National Cancer Advisory Board, it never had any impact on anyone, and in general the doubling in total funds has merely allowed most preexisting progress to become twice as large, thereby not again coming to grips with a real set of priorities.

In fact, the only two new directions taken by the NCI both evolved out of the direct dictates of the 1971 cancer legislation. The first was an expansion of the preexisting center program in the direction of the creation of new Comprehensive Cancer Centers. The models used to justify this expansion were the Memorial-Sloan Kettering complex, the newly created Dana complex in Boston, Roswell Park in Buffalo, and the M.D. Anderson complex in Houston. Twelve new complete centers were to be created to bring the total number to 16. As such, they were to be concerned with all aspects of the cancer problem, both clinical and nonclinical, and therefore to cover the gauntlet from the cause of cancer to the rehabilitation of patients who have been treated for cancer. Unfortunately from the very start, the politics of where these new centers should be located has occupied much more attention than whether their concept really makes that much sense. While clearly no one is against better cancer treatment, it does not necessarily follow that putting the so-called pure scientist under the same organization framework as the cancer clinician will make them collaborate. In fact, it might be wiser to place the best of our scientists working on cancer next to the best of our graduate students, say at Caltech rather than close to the patient at the Los Angeles Cancer Hospital. But for better or worse (I fear much the worse), the 1971 cancer legislation gave heavy legitimatization to the concept of the massive cancer centers. This, despite the fact that the preexisting past history of our comprehensive centers gives us no reason to believe that they would be the best setting for research programs that could be the most difficult that biologists have yet tried to pull off.

But we did not talk about this during meetings of the Cancer Board; we only considered, for example, whether Denver or Seattle made more sense. To be sure in a democracy, there is every reason to fight for the idea of rapidly making medical care equal in all geographical regions. But just like in the war on poverty, the question must be whether we had chosen achievable short-term objectives. It is very hard to create anything of excellence in a short time, particularly massive cancer centers, when there is anything but an oversupply of doctors expert in chemotherapy or the still much more illusive immunotherapy. No matter how hard you try, you can't pull qualified specialists out of a hat. If you don't have deep, deep clinical expertise and imagination, your potential center in Boise, Idaho, for example, will give no better treatment than already provided by its general hospital.

Nonetheless, from the very start the concept of comprehensive centers was a sacred cow; the NCI, responding to the wishes of a large majority of the Cancer Board, set out to bring them into existence as quickly as possible. Essentially the lobbying that produced this climate stated that many more lives could be saved if we set up the centers, and thus being against them was like arguing against motherhood. But I have never heard any good estimate on how many lives could be saved, and so it may be impossible to decide whether they work any better than the best of our local hospitals. Of course, it would be a very different matter if now we had the means to cure the disseminated form of any major cancer (e.g., colon or breast). If that were the case, then those potentially very expensive centers would clearly make sense. But, I fear we don't have an imminent cure, and I suspect that we may be pushing a concept that only can be really successful if completely new ways of treatment emerge. But as of this moment, the "hot leads" are all exceedingly borderline. In any case, I fear that it will be impossible to shut down a comprehensive cancer center even if, at best, it is only marginally more effective than other local hospitals. It will be even more difficult than trying to close down an antiquated VA hospital.

To be sure, doubts were occasionally raised, and I recall at several meetings of the National Cancer Board that Benno Schmidt said we should not go beyond the originally asked for 16 centers. But by now 30 appear to be the target, and I will be surprised if less than 40 exist a decade from now. For it will be difficult, if not impossible, to withhold from a high population center the possession of a highly touted "comprehensive cancer center," even though there may be no statistical yardstick by which their supposed superiority over that of our best non-cancer hospital centers can be proved.

Such sentimental decisions, however, may become increasingly large albatrosses in the aftermaths of the Vietnamese debacle and our oil induced current recession. When the dollars for cancer research cease to grow, will we find that we have created a significant number of first-rate centers for cancer research like the one MIT has just founded, or will we have committed all too much of our money for a hollow facade that cannot be torn down? Alas, if only we ask this question to the doctors in charge of the comprehensive clinical centers, we must expect reports of glowing progress. So it will pay, painful as it may be, to find ways of having such judgments made by people whose future careers are not tightly bound to their continued funding.

The second Congressionally mandated cancer program was that of "Cancer Control"—this was to be a funding device to bring better diagnosis and treatment to the average victim. Initially funded for only twenty-five million dollars, this program will soon grow to some one hundred million dollars, and could easily consume vastly greater sums if happily it can work. But early detection is tricky to bring off, and except for the Pap test is not easily achieved. Now the main new thrust of the control program goes toward breast cancer, a most obvious goal considering its steady pernicious rise and the relatively young age at which it can strike. But it is far from clear whether over a long-term interval you can achieve large scale early detection without turning American women into hopeless nervous wrecks. And the same psychological facts of life urge against the early detection of most other major cancers.

But since "Cancer Control" so directly brings the National Cancer Program to the general public, it seems bound to grow, and very easily could consume a third, if not more, of the total NCI budget. Here we should be very clear that the essence of the program is not research but immediate help to our nation's people. So unless we have some way to accurately judge its success, it could acquire a momentum that could directly depress the money that goes toward understanding, if not preventing, the major cancers. Considering the vast, vast sums probably necessary, if very early detection is ever to be achieved on a national scale, we may find that it makes most sense to spend much of this money to see that known environment carcinogens are kept away from the American public. But such a goal will bring the NCI into direct conflict with very powerful industrial lobbies, and at least when I was on the Board (1972–1974) there was a noted reluctance for the NCI to take on any regulatory role.

On the much more positive side, the new cancer money has permitted about every form of cancer research to get more money. Vital oncology, chemical carcinogenesis, and the biology of the cancer cell, as well as programs for the development of new drugs, are all financially much better off. But here we should note that most of this new money has gone to institutions (companies) already engaged in large-scale cancer research before the passage of the new cancer legislation. The only major exception is the center we are here today dedicating. So essentially we are placing most of our faith—that is money—in the intellectual traditions of the cancer research world that existed prior to the Act. At the same time the nonclinical academically based biological community, such as the part of Harvard I come from, has gotten much poorer in an absolute sense. Yet it has been this academic nonclinical world which has largely been responsible for the amazing development of the new biology of the past 25 years. So we may be witnessing a transference of power (money) from the research-oriented universities (departments) that have made American biology as it now exists, to a new power base whose past existence was derived from its willingness to work on cancer at a time when most scientists thought it to be an intellectual graveyard.

Here we must not solely fault the National Cancer Institute and its advisors. Instead, much blame must also fall on our "leading" pure science institutions (departments) for not realizing that what they are now doing may in the long term contribute more to a cure for a major cancer than any of the so-called relevant research now done in the official cancer centers. So instead of initiating large-scale programs in animal cell biology and asking for commensurate construction funds to house such activities, they have let the major expansion made possible by the conquest of can-

cer legislation fall to those institutions that already had preexisting development (fund raising) offices geared toward cancer and who, at the first
hint of piles of new money, assembled the architects to draw up plans for
the new comprehensive centers. So all too often the NCI has had to place
their money with institutions of doubtful intellectual tradition and created still others in intellectual environments not known to foster bold science.

It would be far better to be slow in the starting of new cancer research
institutes rather than staff them with scientists not at home with the best
of contemporary biology, because unless we wish to place all our faith in
pure luck, we must realize that high-quality cancer research is likely to be
much more difficult to pull off than most other forms of biology. We must
never forget that we are largely in it not because it is the next obvious
objective on an intellectual basis, but because of a desire to alleviate a
major source of human suffering. Now even though the quality of the best
of cancer research is at last on a pace with the best of modern biology, we
must again accept the fact that all too many of the challenging problems
in cancer are at best marginally attackable with today's methodologies. So
a deeper intellectual sparkle may be required for successful cancer research
than for, say, the current challenges of microbial molecular genetics. Yet I
would be less than honest if I said that I thought the average quality of the
American scientist doing cancer research is rapidly going up. At this
moment I fear it is at best holding its own because we have chosen to place
most of the new efforts in institutions who never had much success in
attracting the very best.

So we must not be surprised that for the first time, visible attacks on
the National Cancer Program are beginning to appear. The most damag-
ing of these, by the journalist Daniel Greenberg, has just appeared in the
Washington Post (January 15, 1975) and argues that the American public is
being sold a nasty bill. He points out that even though one in three cancers can now be cured by either surgery or radiation treatment, these
encouraging results were largely obtained prior to 1960 and that despite all
the recent claims about the great potentiality of anticancer drugs, they still
only can cure some 1–2% of disseminated cancers. At the same time, the
statistics for the major cancers (e.g., lung, pancreas, and colon) remain as
gross as ever. Yet this is not the impression that the NCI press office tends
to create, and so Greenberg charges that all too many of their news releas-

es have that ring of honesty that we came to associate with the Vietnamese body counts of the Pentagon. But then Greenberg goes on to weaken his case by stating that we should push more of our cancer money toward new applied approaches like those involving vitamin C and promulgated so vigorously in the press by Linus Pauling and Albert Szent-Gyorgyi. So Greenberg started up with tough unpleasant facts and ended up with nonsense.

The trouble now in the cancer clinic is not that plausible new approaches are being overlooked in favor of past sterility. It is, as I emphasized earlier, that we may not have even one real hot clinical lead that has a sure chance to lead somewhere soon with a major cancer. So we must be much more careful than we have been in the past as to what we allow our lobbyist friends to claim for us. We must not passively sit and listen to people who through overoptimism or premature senility promise things that are very unlikely to happen soon. We have watched over too much of this over the past few years. We should do the science we are trained for and not hold the carrot too close.

We have to sell ourselves, in short, for what we really are, and not for our dreams that the knowledge we now obtain can be immediately applied against some terrible form of cancer. But if we respond to the fear of less cancer money for next year by flashing out even shakier new leads, say in tumor immunology, to mask the fact that we still have not made the big breakthrough, we have nowhere to go but down. Eventually the general public will come to regard the scientific community with governmental press officers. Yet, nevertheless, I fear that as more attacks on the War on Cancer materialize, the NCI may feel it can only defend itself well by asking for still more money. The easiest thing to do when you don't know what to do is ask for more cash. So we must expect that they will be greatly tempted to ask for still more money for still more comprehensive cancer centers and still more cancer control, blaming the lack of clinical progress not on lack of real new ideas or clinical procedures but on the unavailability of money to carry out achievable objectives.

But the best thing I think our country can now do in the War Against Cancer is create just one more institute of high-quality cancer research equivalent to what MIT has started here today. That will do more than all the official hoopla we now receive through the daily mail and press.

Maintaining High-Quality Cancer Research in a Zero-Sum Era

1981

The "War on Cancer" is now a decade old. As with most battles that go on too long, the hoopla and bravado are now muted, and the charismatic generals who were to lead us to certain victory have been rotated to less conspicuous commands. Today we are led by the better staff colonels of the immediate past, seasoned enough not to openly criticize the tactics of yesteryear, knowing well that if their former bosses are perceived not to have been equal to their charges, such lack of confidence may soon be transferred to them as well.

All too clearly the press is testy, and even Congress, old enough to be immediately at risk, cannot be counted on for still another generation of new recruits to march into battle against an enemy that still has not yet revealed its exact shape. Blind bombing for bombing's sake, although emotionally satisfying, was the wrong way to proceed against the Nazi beasts and might well be an equally inappropriate way to bring the cancer cells to their knees. Might it not now be better to tactically retreat to regroup our forces until we can strike with the precision of final victory?

If so, conceivably it could be no catastrophe, but, in fact, a blessing, for our official slasher, Mr. Stockman, to cut back the funding of an NCI that, in growing so fast on the public purse, may have put on fat at the expense of muscle and that, by possibly being so prematurely superfunded, has never acquired the intellectual confidence that comes from the making of real decisions. There was, in fact, an initial OMB thrust to cut away a billion dollars from the NIH (NCI), but it came to naught, and the money instead came from the poor and socially unfortunate. We thus must be aware that given the present national semi-consensus to spend more money on defense, we might only be able to retain our current ways of doing science by taking still more from the unblessed, hoping that the benefits of our research will trickle back to society either through the development of more efficient medical practices or through the creation of the biotechnologic industries in which our nation has a competitive edge.

The questions thus must be faced of not only whether the medically oriented science we now do is really that good, but also whether we are doing it in a properly frugal manner. Are we still all too casually writing purchase orders to consume the vast sums that are popularly thought to have been flowing in our direction ever since Sidney Farber, Mary Laker, and Benno Schmidt coopted first Senator Yarborough, then Teddy Kennedy, and finally Richard Nixon to declare war on cancer. This was to be a campaign fought on many fronts, ranging from the curing of cancer to its prevention. Research as to its nature was to be somewhere in the middle, but not in any real way neglected. Soon we were to have at least a billion per year to spend, and so no one with even the chance of a sound idea need go wanting.

Here I will only comment on the ways we have gone about finding the origin and nature of cancer, since this is the area in which our laboratory has been directly involved. The popular perception now is that we put our money on the wrong horse by not worrying enough about the many chemicals that can cause cancer and instead mounting a large crash program to discover putative human tumor viruses. If viruses were the cause of much cancer and we could develop effective vaccines against the relevant ones, then within a decade or two the incidence of cancer should radically drop. But, after ten years of big funding, not one virus has been implicated as the cause of a major human cancer. Moreover, many experts now increasingly believe that at least 90% of human cancer is due to lifestyle (e.g., smoking, the eating of fat, etc.) and environmental factors. Thus, with ever-increasing hostile scepticism about viruses being significant cancer-causing agents, shouldn't we now, better late than never, deemphasize the various tumor virus efforts? By so doing, the money that now supports them could go, instead, toward mounting a first-rate program on chemical carcinogenesis, a field still badly in need of intellectual feeding.

The truth, however, is not that simple, and, in fact, we are actually way ahead of the game by having directed so much early effort toward the understanding of tumor viruses. It is not that a decade ago chemical carcinogenesis was not inherently important. More relevant was the fact that there was no way this field could efficiently absorb vast sums of new money. It was intellectually in its infancy, and large numbers of new scientists could have been brought into this field only to discover that they had no real working tools. The answers the public wanted could not have been obtained in a reasonable period of time. Now, however, thanks to the new procedures of recombinant DNA, it is at last possible to seek the exact genetic changes brought about by chemical carcinogenesis. Though this will not be an inexpensive endeavor, it is at last the time to back a big intellectual push for the nature of the genes that have become altered through the action of chemical agents.

This, however, should not be accomplished by wide-scale cutbacks in the tumor virus area. To be sure, we do not seem finally about to demonstrate that tumor viruses are, in fact, responsible for major killers such as cancers of the lung or colon. Nonetheless, tumor viruses increasingly hold the center stage in cancer research. Through studies on the molecular level of both those tumor viruses that contain DNA and those that have RNA chromosomes, we are now on the verge of understanding the fundamental chemical features of cancer cells. This happy situation, which a decade ago we thought only marginally possible, is not the result of the work of a few talented scientists. Instead, it is the end product of the moving into tumor virology some ten years ago of a large number of our best younger molecular biologists. Then, recombinant DNA was not even a dream, and the smaller tumor viruses possessed the only eukaryotic DNA (RNA) that seemed simple enough to one day master. Given the bribe of massive funds from the War on Cancer, there really was no other career choice to make. So, within several years, a perceptible increase in the intellectual quality of cancer research was already apparent, and now the snickers that invariably once followed the description of a project as cancer research are gone. By transforming so much of cancer research into an intellectually respectable discipline, the War on Cancer has produced a most-needed victory.

Equally important have been the key research facts that have already come out of our virus efforts. Already the onocogenic segments of a number of tumor viruses, of both the DNA and the RNA classes, have been precisely defined, and we are in the process of clearly establishing the biological function of their protein products. SV40 DNA, for example, codes for two onocogenic proteins, one of which helps to initiate DNA synthesis, the other of still unknown cytoplasmic function. The Rous virus produces sarcomas because it has picked up accidentally a cellular gene *(sarc)* that codes for an enzyme (kinase) that phosphorylates key cellular proteins. When present on the viral chromosome, the gene is not subject to normal cellular control and overproduces its kinase products. The overphosphorylation leads, in ways that we have yet to discover, to the sarcoma phenotype.

Initially, the pace at which we could work on these vital cancer (onco) genes was of necessity both slow and expensive, with much luck being necessary just to obtain the mutant virus strains needed. Now, however, with recombinant DNA, the pace of tumor virus research has quickened almost frighteningly, and the normal cellular roles of the "cancer genes" carried by many more tumor viruses are likely soon to be known. With them in hand, we shall have an extraordinarily powerful set of tools to firmly establish the basic biochemistry of cancer.

The last stages of this victory, however, will not be a walkover. In moving our prime molecular attention from the tumor viruses to the cancer cells themselves, we face an extraordinary increase in complexity. The simplest human cells contain at a minimum some 2000 different proteins, the interactions of many of which must be dissected before we can be sure we have the right answers. The equipment we shall have to use will of necessity be more sophisticated than that employed in the past and, unless we are much luckier than we deserve, will not be cheap.

So, I don't foresee any way that we can maintain our momentum, much less increase it to the extent now technologically possible, if the NCI monies available for these fundamental studies are even mildly slashed. In a real sense, we are victims of our enormous success. Until recombinant DNA came along, we accepted the fact that all the prime parameters of cancer cells might not be in for many decades to come. Now with the smell of victory in the air, we see every reason to push ahead as fast as possible. This pursuit will take more, not less, money at a time when even the maintenance of our status quo is bound to strike many outsiders as basically indecent.

Moreover, soon we may be approaching the time when it actually makes sense to spend substantial monies looking for human tumor viruses. As we finally begin to master the basic molecular biology of the RNA tumor viruses (retroviruses), the more obvious it becomes that those scientists who, at the beginning of the War on Cancer, staked their careers on implicating retroviruses as human carcinogens never had a reasonable chance of success. They did not know enough or have the technological handles to proceed sensibly. Now that we appreciate the transposonlike qualities of retroviruses, we at last have a philosophy, admittedly incomplete, for analyzing the various retroviruslike entities that will be found over the next few years in a variety of human cells. By cloning them in bacteria, we can easily test whether or not they contain cellular (onc?) insertions into the basic retroviral genome. Even if we find "onc-like" segments, however, and then through transfection experiments show that they are real "oncs," we shall not yet have our "smoking guns."

To find them, we might best concentrate less effort on the isolation of new retroviruses and more on directly testing the DNA from primary human tumors to see whether it can transform the appropriate normal cells into their cancerous equivalents. Preliminary attempts have already given positive results, and we have every reason to expect that a number of human "cancer genes" will soon be cloned in bacteria. Then, through DNA sequencing studies, we can find out whether any of the cancer genes are closely associated with retroviruslike transposons. If any such putative retrovirus sequences are identified, it should be relatively simple to look next for their expression in various human tissues as the first step to seeing whether in their RNA form they are horizontally transmitted from one cell to another as infectious retroviruses. Those experiments, at long last within our techological capability, will not be cheap. So unless there is the prospect of ample long-term funding for such efforts, we shall remain ignorant longer than necessary as to the role of retroviruses in human cancer.

The time may also soon be ripe for looking much more closely at the various papilloma (wart) viruses, about which there is no doubt as to their proliferative potential. Because they still have not yet been successfully grown in cultured cells, they were of necessity refractory to molecular analysis until recombinant DNA came along. Now we have the potential to routinely screen hundreds of human tumors for the presence of papilloma DNA. Definitive answers should also now be possible as to the oncogenic role, if any, of the various herpes viruses and in particular of EB, which has long been identified with the proliferative aspect of mononucleosis and thus always a tantalizing candidate as the causative agent of one or more of the Hodgkin's-disease-like syndromes.

Even if it turns out that viruses cause only a minor fraction of human cancer, it still makes great sense to try to obtain the respective vaccines. The doing away with any minor cancer by such procedures would in a decade or so more than pay for all the monies so far spent on tumor virus research. So, given the availability of funds, we would be most remiss to still be held in check by our lingering embarrassment over the premature excesses of past human tumor virus searches. Instead, we should confidently take advantage of the fact that soon we shall have the appropriate base of DNA science to let us proceed in a logical, sensible manner.

Here we must not lose sight of the fact that a main reason why we are excited about the next decade in cancer research is that at last it has attracted many of the best younger scientists. If, however, the cancer money targeted toward the molecular genetics approach noticeably decreases, all too many of the better young postdocs now working with viral oncogenes will see the writing on the walls and move elsewhere. We are now out of the prerecombinant DNA days when tumor virus nucleic acids provided the best way to get at the eukaryotic genome. Tumor virology at the molecular level is no longer the only place to go, independent of whether or not you are interested in cancer. It is only one of many aspects of molecular biology that is intellectually exploding because of the availability of the marvelous new gene splicing tricks.

We also must face the increased competition for bright young brains that is already here from the creation of the many new recombinant DNA companies as well as from the major chemical giants that now privately proclaim that they will only have a long-term future if they successfully embrace biotechnology. Therefore, we cannot take for granted that the high intellectual standards that now characterize viral oncology will be maintained over the next decade.

It would thus be a tragic mistake now to rebalance money away from the virus area in order to help protect less-strong facets of the NCI program from the ravages of inflation and mounting entitlements. The effect on NCI could easily be that of a partial lobotomy.

I therefore see no choice now but for the NCI and its NIH parent to look much more closely at its various decisions to see which programs really are effective and which are really there only because no one wants to blow the whistle on good intentions gone astray. Here it is important to emphasize that, except for the NCI, we have been painfully shedding the fat from our individual NIH research grants ever since the last days of Lyndon Johnson; and for the last several years, the same has been true for the NCI. Significant further retrenchments in the size of individual grants will most certainly be at the expense of quality. The average salary paid academic scientists has not kept pace with inflation, and we have reached the state where either two bread winners per family or some moonlighting has become a prerequisite for a modestly adequate living. We are, furthermore, suffering not only from the use of equipment past its prime, but also from having to live with falling supply budgets, which soon must lead to a cutback in the number of good experiments. At least in this laboratory, I don't see much room for additional supply savings that will not compromise our productivity, and my impression is that a similar situation exists in most other leading institutions. Almost everywhere the general picture seems to be of overruns in supply budgets that soon must lead to actual lab shutdowns until the deficits are paid back.

Thus, the future fostering of high-quality research on cancer now requires the serious pruning away of the not quite best. How to distinguish the best from the almost best, however, will never be easy, and in doing so we are bound to make some wrong choices and cut individuals out of scientific careers that they have been given every indication to expect. Wielding this particular axe will be a most unpleasant task, but it is a job that the NIH (NCI) directors can no longer avoid. As long as we were living on growing budgets, the hard choices could be avoided. Now, either reaffirming the importance of tumor virus research and keeping up its magnificent upward crescendo or upgrading the equally important area of chemical carcinogenesis research cannot be passive deeds. Such acts can only be done by deciding that other aspects of cancer research have lower chances of producing home runs, and such pronouncements are bound to offend many long-time, powerful supporters of cancer research. But, as I already know too well, the function of a director is to lead, not follow, and we can only hope that the leaders of NCI (NIH) and their counselors will do their jobs well.

The Science for Beating Down Cancer

1990

The understanding of cancer as an aberration of the normal processes of cell growth and division has long stood out as a prime, if not
ultimate, goal of the world's biomedical research community.
When Everest was conquered, we saw the challenge of cancer in terms of
that long-sought-after "Himalayan" goal. Now, with time, we see that we
are assaulting a more-K2-like peak, where the ice falls ahead pose complexities that even the resolute and strong know may be close to the limits
of human endurance.

When I was a boy growing up in Chicago, cancer was only talked
about in whispers, a scourge that struck at random and against which we
had no medical means of fighting back, particularly if it had spread
beyond its point of origin. The only hope had to be the new facts that science would one day discover. After I became a student at the University of
Chicago in 1943, I became aware of the research monies given to its Medical School by Albert and Mary Lasker. They wanted the world to have the
scientific knowledge that would reveal the real faces of the enemy. Blindly
thrashing against a foe that we knew only by name and not by form or substance could only perpetuate our fears. So, the Laskers put real life into the
body known as the American Society for Cancer Control, then an assembly of doctors, many of whom were habituated to keeping the truth from
their patients. Their efforts transformed this ineffective organization into
the American Cancer Society, a national body founded in 1944 that would
work for the public good by using the scientific mind, as opposed to the
surgeon's scalpel, as the means ultimately to banish cancer from the
human vocabulary.

It was in this same year that Avery, MacLeod, and McCarty published their historic paper showing that DNA molecules, not proteins, were the hereditary molecules of bacteria. At that time, very few scientists worked on or were excited by DNA or with RNA, its equally mysterious companion nucleic acid. So the Avery result did not immediately galvanize a then-unfocused biological community into a DNA-dominated mentality. I heard first of Avery's experiments through Sewell Wright's course on physiological genetics that I attended in the spring of 1946, my junior year in college. DNA's potential significance, however, only hit me with a vengeance when I moved on in the fall to Indiana University as a graduate student. There, my first-term courses brought me into the center of the gene replication dilemma: How could genetic molecules with their very great specificities be exactly copied?

Salvador Luria in his virus course excited me about these still very mysterious disease-causing agents, while Hermann J. Muller in his advanced genetics course recounted his lifelong odyssey in search of the secrets of the gene. The simplicity of Luria's bacterial viruses (phages) immediately fascinated me, and in the spring of 1949, I started my Ph.D. thesis research in his lab. A year earlier, Luria had become one of the first recipients of the American Cancer Society's (ACS) new research grants. These were very important monies for the fledgling scientific area that later was to be known as molecular biology. The National Science Foundation (NSF) had not yet come into existence and the National Institutes of Health (NIH), much less its National Cancer Institute (NCI) component, had only a minuscule budget for scientific investigators outside its own Bethesda walls. At that time, the ACS did not have its own staff for evaluating grant proposals, a task then given to the Committee on Growth of the National Research Council.

Support of Luria's research by the ACS was not at all surprising. Several types of viruses were known to cause cancer in a variety of animals, and it was natural to think that some human cancers might also have viral etiologies. But in those early postwar years, no one really knew what a virus was, and directly attacking how the cancer-inducing viruses acted was not a realistic objective. Many viruses, including Luria's little phages, contained DNA, and we often speculated that DNA was their genetic component. But there were other viruses that totally lacked DNA and instead had RNA components. Conceivably, both forms of nucleic acid carried genetic information, but there still existed many scientists who suspected that maybe neither DNA nor RNA was a truly genetic molecule. Perhaps there was a fatal flaw to the Avery experiment that no one had yet caught.

General acceptance of the primary genetic role of DNA only came when Francis Crick and I found the double helix. The fact that it had a structure that was so perfect for its self-replication could not be a matter of chance. When I first publicly presented the double helix at the 1953 Cold Spring Harbor Symposium, there was virtually immediate and universal acceptance of its implications. At long last, we had the reference molecule on which to base our thinking about how living cells operate at the molecular level. And the DNA viruses immediately could be viewed in a more approachable fashion. The DNA molecules within them were clearly their chromosomes; and within a year, Seymour Benzer produced a detailed genetic map of the T2 gene of phage T4 in which the mutant sites along a gene were correctly postulated to be the successive base pairs along the double helix.

Less obvious was whether RNA also could be a genetic molecule. By then, we believed that RNA functioned as an informational intermediate in the transfer of genetic information from the base-pair sequences of DNA to the amino acid sequences in proteins (DNA→RNA→protein). As such, RNA did not need to be capable of self-replication, it only needed to be made on DNA templates. Vigorous proof that RNA also could be genetic molecules came from Alfred Gierer's 1956 demonstration in Tübingen that RNA purified from tobacco mosaic virus was infectious.

Viruses by then had become perceived as tiny pieces of genetic material surrounded by protective coats made up of protein (and sometimes lipid) molecules that ensured their successful passage from one cell to another. At first, we believed that the proteins used to construct their outer coats were the only proteins coded by the viral chromosomes. The enzymes used to replicate their DNA were initially believed to be those of their host cells. This assumption, however, created the dilemma of how the RNA of RNA viruses was replicated if the RNA within their host cells was made entirely on DNA templates. One way out of the dilemma was to postulate that some types of DNA-made RNA were later selectively amplified by RNA-templated RNA synthesis catalyzed by host cell enzymes. By 1959, however, Seymour Cohen's and Arthur Kornberg's labs began to show that phage chromosomes coded for many of the enzymes needed to replicate their DNA. This opened up the possibility that animal viruses also coded for the enzymes that duplicate their DNA. If true, I thought we might have the first real clue as to how viruses cause cancer.

Dissolving the deep enigma surrounding viral carcinogenesis first became a goal of mine when I learned of tumor viruses in Luria's 1947 virus course. A young uncle of mine was then dying of cancer, and it was that fall when I first acutely sensed the need for science to fight back. This desire was rekindled in the spring of 1958 when I was visiting Luria's lab, by then in Urbana. There I heard Van Potter from the University of Wisconsin, Madison, give a lecture on the biochemistry of cancer. From him I first realized that the cells of higher organisms, unlike those of bacteria, needed specific signals to divide. In fact, the majority of cells in our bodies are not dividing but are in an apparent resting state where DNA synthesis is not occurring. A DNA virus infecting such cells would be unable to multiply unless it coded for one or more enzymes that specifically functioned to move their quiescent host cells into the "S" phase of DNA synthesis. Conceivably, insertion of animal viral genomes into the chromosomes of resting host cells would convert them into dividing cells with the signal for DNA synthesis always turned on.

This idea came to me when I was preparing a lecture on cancer to beginning Harvard students whom I was trying to excite with the new triumphs of molecular biology. It was they who first learned of my hypothesis. That viral carcinogenesis might have such a simple answer dominated my thoughts all that spring of 1959. In May, I presented it as my prize lecture for that year's Warren Award of the Massachusetts General Hospital. Francis Crick and I shared the award, and my presentation came after he told an overflow audience how transfer RNA was the "adaptor" molecule he had earlier postulated for reading the messages of RNA templates. Thanks to Mahlon Hoagland and Paul Zamecnik's new experiments, there were no doubts as to whether Francis' ideas were on track. His talk had the virtue of being not only elegant, but also right. On the other hand, my talk had to seem more hot air than future truth. I left the Museum of Science Lecture Hall depressed at the thought that I had appeared at least an order of magnitude less intellectually powerful than Francis. Clearly, I might have given a more convincing talk if I had a plausible hypothesis as to why RNA viruses also sometimes induce cancer. As opposed to the situation with DNA, resting animal cells are constantly making RNA. Conceivably,

there were two very different mechanisms through which the DNA and RNA viruses caused cancer. The other possibility was that my idea, although pretty, was just wrong.

During that spring, I started some experiments with John Littlefield, who had been purifying the Shope papilloma virus from rabbit warts. At that time, it was the smallest known DNA tumor virus and we expected its DNA to have a molecular weight of about four million. Surprisingly, using sedimentation analysis, we measured an apparent molecular weight of some seven million, with some molecules of seemingly twice that size that we thought might be end-to-end dimers. Unfortunately, we never looked at them in the electron microscope. If we had, we would have discovered that the Shope papilloma DNA is circular and that the faster-sedimenting molecules were not dimers, but a supercoiled form of the uncoiled simple circle that has a molecular weight of five million. In retrospect, I felt stupid, because earlier I had spent much time arranging for Harvard's Biological Laboratories to get an electron microscope. But, circles were not yet in the air and I never expected to see anything interesting.

Then the only, and not always dependable, source of the Shope papilloma virus was the Kansas trapper Earl Johnson. So, the discovery of a more accessible and even smaller mouse virus was beginning to revolutionize tumor virology. In 1958, Sarah Stewart and Bernice Eddy, working at the National Cancer Institute, opened up the DNA tumor virus field to modern virological methods through being able to propagate Ludwik Gross' "paratoid" virus[1] in mouse cells growing in cell culture. They renamed this virus "polyoma," after its unexpected property of inducing a broad spectrum of tumors following inoculation into immunologically immature newborn mice. Quickly, several molecularly oriented virologists, including Renato Dulbecco at Caltech, Leo Sachs in Israel, and Michael Stoker in Glasgow, took up the polyoma virus system, moving on several years later to the newly discovered, similarly sized monkey tumor

[1]Ludwik Gross is a Polish-born cancer researcher who discovered the RNA-containing mouse leukemia virus without research support, while working as a full-time hospital doctor in New York. His results ran counter to then prevailing orthodoxy and were initially widely held to be erroneous, even fraudulent.

virus SV40. Easily workable cell culture systems to study the multiplication of these viruses as well as their cancer-inducing (transforming) properties were in place by the early 1960s. The time had thus come to ask whether these DNA viruses contained one or more specific cancer-inducing genes. So, there was an aura of real excitement permeating our 1962 Symposium on "Animal Viruses." I came down from Harvard and listened to Dulbecco give the closing summary at which the then-young Howard Temin and David Baltimore were much in evidence.

I did not, however, then join in the cancer gene quest. My earlier experiments on papilloma DNA were diversions from a ribosome-dominated lab that unfortunately was still in the dark as to where to go next. But when we found the first firm evidence for messenger RNA in March of 1960, the course of my Harvard lab for the next decade was firmly set. We wanted to understand how messenger RNA was made and then functioned to order the amino acids on ribosomes during protein synthesis. But I continued to emphasize tumor viruses in my Harvard lectures, which eventually formed the basis for my first book (1965). Its last chapter, "A Geneticist's View of Cancer," discussed the DNA and RNA tumor viruses, concluding with the statement that through study of the simple DNA and RNA tumor viruses, we have our best chance of understanding cancer.

My taking on the Directorship here early in 1968 at last gave me the opportunity to get into the DNA tumor virus field. The great era of phage and bacterial genetics research at Cold Spring Harbor was coming to an end and we needed a new intellectual focus as a reason for our existence. Fortunately, that summer, our animal virus course brought to us several individuals who were just starting research with tumor viruses. The lecturer to excite me most was Joe Sambrook, then a postdoc in Dulbecco's Salk Institute lab where he worked with Henry Westphal on the integration of SV40 DNA into the chromosomal DNA of cells made cancerous by SV40. Soon after meeting Joe, I asked him to move here and start a DNA tumor virus lab. He quickly accepted, wrote a successful grant application, and arrived here the following June.

Within several years, Joe was leading a very high-powered group in James lab consisting of Henry Westphal, Carel Mulder, Phil Sharp, Walter Keller, and Mike Botchan whose main purpose was to identify the SV40

gene(s?) that leads to cancer. By then, George Todaro and Robert Huebner, members of the NIH Special Cancer Virus Program, proposed using
the name "oncogene" for a cancer-causing gene, and this designation
rapidly caught on. Through work here and at several other key sites,
including the Salk Institute, NIH, and the Weizmann Institute, the SV40
oncogene(s?) was shown to be identical to the so-called "early gene(s)"
that functions at the start of the SV40 replication cycle. This was a most
gratifying result, compatible with my brainstorm of a decade earlier that
the DNA viral oncogenes function to convert host cells into states capable
of supporting DNA synthesis.

Key tools in everyone's analysis were the newly discovered restriction
enzymes that cut DNA molecules at precise nucleotide sequences. A given
restriction enzyme was used to cut up a viral genome into discrete pieces
that could then be isolated from each other by a powerful new ethidium
bromide agarose gel procedure developed here by Phil Sharp, Bill Sugden,
and Joe Sambrook. Restriction enzymes came to the laboratory when
Carel Mulder brought Herb Boyer's RI into the James lab and later went
on to isolate the SmaI enzyme. Soon afterward, Joe Sambrook and Phil
Sharp began to use HpaI and HpaII enzymes, whose use was pioneered by
Ham Smith and Dan Nathans at Johns Hopkins. More and more restriction enzymes came on line through the efforts of Rich Roberts, who
arrived here in late 1972. About 50% of the world's commonly used
restriction enzymes were discovered over the next decade in the Roberts
laboratory.

By then, James lab was also working with a second DNA tumor virus,
a human adenovirus that Ulf Pettersson had brought from Uppsala in
1971. Its life cycle also was divided into an early phase and a late phase,
with the genes carrying its oncogenic potential also being "early" genes. A
clear next objective was to find and identify the messenger RNAs for proteins encoded by the early and late adenovirus genes. Ray Gesteland's and
Rich Roberts' groups in Demerec lab took on this task, which soon began
to generate mystifying results. All of the late mRNAs seemed to possess a
common terminal segment even though they were encoded by widely separated sequences of DNA. Resolution of the paradox occurred in late
March of 1977, when Rich Roberts orchestrated a team consisting of Tom
Broker, Louise Chow, Rich Gelinas, and Dan Klessig to their monumental

discovery of RNA splicing. Independently, Phil Sharp and Susan Berget, then at MIT, made the same great discovery after initially observing with the electron microscope that the 5′ end of a late adenovirus messenger RNA did not behave as expected.[2]

The discovery of RNA splicing was a once-in-a-lifetime event that completely transformed all of eukaryotic biology, and our 1977 Symposium, where the discovery was first publicly announced, was an occasion of intense intellectual ferment. Afterwards, new implications arose virtually weekly. Among the first was understanding that the T(umor)-antigen-coding SV40 early gene specified two different cancer-causing proteins. They are derived by two different ways of splicing the early SV40 messenger RNA. The once-thought single SV40 T(umor) antigen in fact consists of large (T) and small (t) components, with the main cancer-causing activity due to the large T antigen. Splicing also occurs with the early adenovirus mRNAs, with two of the resulting protein products of the E1A and E1B genes playing essential roles in early viral replication as well as having oncogenic activity.

Now we realize that the subsequent working out of how these tumor virus oncogenes actually induce cancer would have been virtually impossible if the procedures of recombinant DNA had not been discovered in 1973. They have allowed us to study the action of individual oncogenes as well as to prepare the large amounts of highly purified oncogenic proteins needed to study their molecular functioning. We had, however, to wait six long years after Herb Boyer and Stanley Cohen gave us the first generally applicable recombinant DNA procedures until the stringent NIH prohibitions against using recombinant DNA to clone viral oncogenes were dropped. Only in early 1979 could the recombinant DNA era of tumor virology take off.

Not only were the DNA tumor viruses ripe for analysis, but how to think about the RNA tumor viruses was also known. Through work in the 1960s by Harry Rubin, Peter Vogt, and Howard Temin, the defective nature of most RNA tumor viruses had been firmly established. Their replication

---

[2]Splicing is an important feature of gene expression. Most genes (except in bacteria) contain regions, called introns, which do not encode any of the sequence of the protein that the gene specifies. After the DNA has been copied into messenger RNA, these regions are excised by special enzymes. The resultant pieces of RNA are then joined up (spliced) by other enzymes to form the functional messenger. The pieces of RNA are sometimes joined together in varying order and thus give rise to a family of related proteins (splice variants).

requires the simultaneous presence of a normal helper virus. In acquiring their cancerous potentials, the RNA tumor viruses had somehow lost part of their own genomes. The way such RNA genomes, be they normal or defective, are replicated was first correctly hypothesized by Howard Temin. In 1964, he suggested that the infecting RNA molecules served as templates to make DNA genomes, which, in turn, integrated as proviruses into host cell chromosomes. Proof came in 1970 when Howard Temin working with Satoshi Mizutani, and independently David Baltimore, discovered within mature RNA tumor virus particles the enzyme reverse transcriptase. Soon the name "retroviruses" became used to encompass all those RNA viruses that replicate their RNA through a DNA intermediate.

Studies on the Rous sarcoma virus (RSV) provided the first deep insights on how the RNA tumor viruses cause cancer. RSV mutants that were unable to transform cells were found by Peter Duesberg and his collaborators to frequently lose part of a specific region of the genome that they called *src*. Its true nature became known from Mike Bishop's and Harold Varmus' 1976 seminal experiments showing that the sequences within RSV are highly homologous to those of a normal cellular equivalent. This finding immediately suggested that the cancer-causing signals of retroviral genomes have nothing to do with the replication processes of retroviruses. Instead, they originated from illegitimate recombinant events that replaced normal retroviral base pairs with DNA segments bearing cellular genes. A year later, the protein product of itself was isolated and found independently by Ray Erickson and Art Levinson to be a protein kinase, an enzyme capable of adding phosphate groups to preexisting proteins.[3]

Over the next decade, more than 30 additional oncogenes were isolated from RNA tumor viruses. In each case, they closely resemble a normal cellular gene. The proteins these oncogenes encode have seemingly very diverse roles; some are growth factors, others are receptors, many are kinases, and still others code for proteins that bind to DNA and control transcription. Unifying the roles of all these oncogenes, as well as of their normal cellular equivalents, is their involvement in the signal transduction

---

[3]Kinases are enzymes that introduce phosphate groups ($-PO_3$) into specified points in proteins. Such phosphorylation, as it is called, can engender large changes in the properties and activity of the protein and is an important factor in control of metabolism.

processes that control whether a cell divides, remains quiescent, or becomes terminally differentiated. In normal cells,[4] the functioning of these signal transduction genes is tightly regulated, so that they function only when cell division is needed. In contrast, the functioning of their oncogene derivatives is unregulated and leads to overexpression of their respective protein products. Proto-oncogene is the term now used to designate normal genes that can be converted into oncogenes by mutations or abnormal recombinational events.

We thus see that the oncogenes of DNA and RNA tumor viruses work in fundamentally different fashions. Those of DNA tumor viruses play essential roles in the replication of the viral genomes, with their cancer-causing attributes related to the tricks by which they turn their normal quiescent host cells into factories primed for DNA synthesis. In contrast, the oncogenes of retroviruses play no role in their replication, having arisen by genetic accidents that dissociate growth-promoting genes from their normal regulatory signals.

When they were first found, the question had to be faced whether the oncogenes of retroviruses were essentially laboratory artifacts and not related to human cancer. One way to settle the matter was to devise procedures that directly looked for human oncogenes by asking whether DNA isolated from human cancer cells could convert a normal cell into a cancerous cell. At MIT, Bob Weinberg first convincingly showed that this could be done using the DNA infection (transfection) procedures that Mike Wigler helped to develop while he was a graduate student at Columbia University. By that time, Mike had joined our staff and was focusing on ways to clone the genes that his transfection procedures had functionally introduced into cells. In 1981, Wigler and Weinberg, working with the same cancer cell line, used different cloning procedures to isolate the first known human oncogene. More importantly, this bladder cell oncogene turned out to be virtually identical to the viral oncogene *ras* isolated sev-

---

[4]Signal transduction is the process by which an event at the cell exterior, most often the binding of a molecule (hormones, such as insulin, are familiar examples) to a receptor on the cell surface, sets in train reactions within the cell. The attachment of the triggering substance to its receptor is pictured as releasing a signal on the inside of the cells to start the required sequence of events.

eral years before by Ed Scolnick when he was at NCI. It was just the first of several retroviral oncogenes shown to be a cause of human cancer. Now no one doubts that the study of retroviral oncogenes bears directly on the understanding of human cancer.

The NIH-3T3 cultured mouse cells that Wigler and Weinberg made cancerous by the addition of the oncogenes were later found to be more predisposed to cancer than cells obtained from the organs of living animals. Now we realize that Weinberg's and Wigler's classic experiments would have failed if they had used cells whose growth regulation was more normal. Two years later in 1983, Earl Ruley, here at James lab, showed that the ras oncogene only transforms normal rat kidney cells to a cancerous state when a second oncogene is simultaneously added. In his experiments, either the adenovirus E1A oncogene or the retroviral *myc* oncogene could complement the activity of ras. At the same time, Helmut Land and Luis Parada working at MIT in Weinberg's lab independently came to the same conclusion: Normal cells do not become fully cancerous through acquiring a single oncogene, but instead they become cancerous progressively, as other oncogenes successively come into action.

Upon arriving at Cold Spring Harbor, Ed Harlow focused on the key adenovirus oncogenic protein E1A to see which proteins it bound. He hoped that by identifying the cellular components with which it interacted, clues would emerge as to how E1A caused cancer. To spot these cellular proteins, he used a monoclonal antibody against E1A to precipitate it from extracts of adenovirus-infected cells. Then, he displayed the resulting precipitate on a gel to see whether proteins in addition to E1A could be detected. In this way, several unknown cellular proteins were found to bind to E1A. What any of them did remained a mystery until in the fall of 1987 a paper was published by Wen-Hwa Lee on the properties of Rb, a newly identified DNA binding protein that helps to prevent cancer. When Rb is absent in a human due to mutations in both of his two genes, retinoblastoma (cancer of the retina) develops. Those individuals who inherit a bad gene from one of their parents are at risk for this cancer, frequently developing retinoblastoma at an early age when a cancer-causing mutation occurs in the remaining good gene. Harlow's lab noticed that the Rb protein had been assigned a size very similar to one of the proteins (105K) that binds tightly to the E1A protein. Hoping that they might be the same, Harlow's lab began a collaboration with Bob Weinberg's lab, one of the three groups that had just cloned the gene. Happily, the two proteins (Rb and 105K) proved to be identical, suggesting that E1A's oncogene potential lies partially in its ability to neutralize the anticancer activity of the Rb protein. I say partially because besides binding to Rb, E1A binds also to several other, still to be functionally identified, molecules, each of which also may be a cancer-preventing (anti-oncogene) protein.

The implication of Harlow's discovery widened with the subsequent finding that the T antigens of SV40 and polyoma also tightly bind Rb, as does an oncogenic protein coded by a papilloma virus that causes warts. Moreover, the once mysterious p53 protein that binds to the SV40 (polyoma) T antigen was shown last year by Arnold Levine and Bert Vogelstein also to be an anti-oncogenic protein. Although p53 does not bind to E1A, it does bind to E1B, a second "early" adenovirus protein that potentiates the oncogenic transformation potential of the E1A protein.

The study of tumor viruses has thus advanced fundamental cancer research more than anyone could have predicted when serious research on them began some three decades ago. The RNA tumor viruses have revealed almost all of the oncogenes known today. Without them, we would largely be in the dark as to the molecular players in the signaling processes that lead to cell division. Equally important have been the insights gained from learning how the oncogenes of the DNA tumor viruses work. The knowledge that they prevent anti-oncogenes from functioning gives us a new way to identify anti-oncogenes and will materially advance our understanding of hereditary predispositions to cancer.

We must remember, however, that the small hyperplastic tumors that result from gain of oncogenes that promote cell growth and division are generally benign. These usually tiny tumors generally only grow to life-threatening size when they become infiltrated (vascularized) by newly growing blood vessels that bring to them the oxygen and nutrients needed for their growth. Douglas Hanahan elegantly showed this through experiments done in collaboration with Judah Folkman. In our Harris laboratory, Doug introduced oncogenes into the germ cells of mice to produce transgenic mice in which these cell-division-signaling genes are expressed in early development as well as throughout adult life. His targeting of the SV40 T antigen to function in the pancreas resulted in large numbers of benign hyperplasmic growths in the insulin-producing islets. Only a small percentage of these benign growths, however, turned into rapidly growing tumors with infiltrating blood vessels. They did so by acquiring the ability to send out angiogenic (blood-vessel-forming) signals that induced neighboring endothelial cells to form new blood vessels. Finding out which molecules carry these signals will soon become a major objective for the coming decade of cancer research. If we could find inhibitors of these angiogenic growth factors, we might have in hand a powerful new way to stop cancers from growing. Most importantly, the endothelial cells that line our blood vessels effectively do not divide in adult life. So, as Judah Folkman has long dreamed, inhibitors of tumor angiogenesis would not necessarily affect the healthy functioning of our preexisting blood vessels.

The truly cancerous cells of solid tumors (as opposed to those of the circulating cells of the blood) also show failure of their normal cellular affinities and spread (metastasize) to many unwanted tissue sites. Our first molecular clue as to the oncogenic changes that create cells capable of metastasis came this past January from the cloning of an oncogene on human chromosome 18 that leads to a final step in the progression of normal colon cells into their highly malignant equivalents. Through a very difficult feat of gene cloning, Bert Vogelstein found that this oncogene codes for a cell membrane protein involved in cell–cell recognition. Over the next few years, the losses of many additional cell-recognition proteins are likely to be implicated in the unwanted spread of tumor cells from their original sites of origin.

At long last, we may thus have the proper intellectual framework to understand most of the more common life-threatening cancers. At their essence are three types of genetic changes: (1) those that make cells divide when they should not, (2) those that lead to the formation of blood vessels that infiltrate into and thereby nourish the growing tumors, and (3) those that modify cell-recognizing molecules in ways leading to losses of their respective cell's ability to recognize their normal cellular partners. We have indeed come a very long way, but there remain myriad further details to unravel about both currently known oncogenes and the many more oncogenes yet to be discovered. Many of these new observations will initially unsettle us and momentarily make us despair of ever being able to have a fair fight against an enemy that so constantly changes the face it presents. But now is most certainly not the moment to lose faith in our ability to triumph over the inherent complexity that underlies the existence of the living state.

Given enough time, and the financial and moral resources that will let those born optimistic stay that way, the odds for eventual success in beating down cancer are on our side.

Societal Implications of the Human Genome Project

Moving on to Human DNA

1986

That our forthcoming 51st Cold Spring Harbor Symposium on Quantitative Biology will focus on the molecular biology of *Homo sapiens* indicates the extraordinary advances now resulting from the application of molecular biology to human beings. As with most new major technological breakthroughs, however, these advances are accompanied by social, economic, and moral implications that should not be ignored. In particular, we must begin to consider the consequences of our newly found abilities to analyze human DNA molecules, capabilities that over the next several decades should witness the working out of the complete sequence of the some 3×10^9 base pairs that comprise the human genome.

Now we are just beginning to see the details of how we humans, as well as other forms of life, are the products of random evolution of DNA molecules that began when the first forms of life emerged in an early phase of the earth's existence some four billion years ago. Then, as now, the various enzymatic steps through which DNA is replicated were constantly generating a broad range of new DNA molecules. These mutant DNAs in turn created genetic variants, some better adapted for existence on the earth as it now exists and others terminal misfits, unable to survive under any conditions. This constant production of new variability is itself under genetic control and now expresses itself most strongly among the larger higher plants and animals whose generation times are relatively long and whose populations are relatively small. Such species have to possess a high degree of inherent variants so that even in times of reduced numbers they contain sufficient genetic variants to survive when confronted with new forms of environmental stress. Humans are thus extraordinarily polymorphic

163

animals, with each individual (except for identical twins) being highly unique, not only at the level of external appearance but also at the level of the structure of their individual genes.

Already it is clear that each of us has easily computerizable DNA fingerprints that distinguish us from all other individuals and that, for example, can unambiguously establish our hereditary relationship to others. Because each of us possesses perhaps some hundred thousand different genes, it is clear that none of us possesses only "good" examples of these genes. Each of us contains a mixture of genes, some of which help greatly to favor our existence and reproduction; others, perhaps the large majority, are evolutionarily neutral, and still others diminish our chances to have functionally meaningful lives. In some cases, a given gene will always be bad, but in many cases the final effects will depend upon where and how we live. Those genes, for example, that bring about light-color skin help to promote necessary vitamin D synthesis in the skin of those who live in colder northern regions but promote skin cancer for those individuals who live near the equator.

Although initially it was possible to talk about good and bad genes largely only in the abstract, the arrival of recombinant DNA procedures, in particular the use of DNA polymorphisms in gene mapping, has transformed the field of human genetics into a virtual beehive of activity. Each week more and more genes are being mapped to precise chromosomal locations, and long-intractable genetic problems such as epilepsy, schizophrenia, and manic–depressive syndromes may be resolved over the next decade. Even more important, many of the genes behind the already well-defined genetic diseases like sickle-cell anemia are being isolated and their protein products identified. Once we have the culprit proteins, we shall be able to zoom in much more effectively on the pathogenesis of still very mysterious diseases like cystic fibrosis. Equally important, once a bad gene has been isolated, tests can be developed to diagnose its presence in newly developing fetuses. Antenatal diagnosis, for example, will soon almost routinely reveal whether a given fetus is doomed to develop into a child with muscular dystrophy or give the reassuring good news that the respective bad gene is lacking and that the arrival of a healthy baby should be anticipated. Thus the possibility will soon be at hand of greatly reducing the long-term social and financial burdens placed on society by genetic diseases, particularly those manifested in childhood.

In accepting genetic disease as an inevitable consequence of human evolution, we are accepting the premise that the human beings who exist on earth today are not the result of any predetermined plan that has given human beings a manifest destiny to dominate the earth. Instead, our unique capabilities as humans come from the evolutionary events that have allowed our species to develop our very powerful brains. It is this specialized collection of nerve cells that allows us to be so conscious of the outside world, memorizing some of this information and later selectively recalling some of these facts as we think. None of these cognitive attributes by themselves, however, are unique to humans. There is every reason to believe that many higher animals think in some ways like humans, possessing senses of beauty, having friends and enemies, and being anxious, happy, or sad, depending upon how their respective lives go. What makes us so remarkable as higher animals is not our ability to perceive pleasure or pain and act accordingly but our capacity to develop highly structured languages both spoken and written. Through the use of these languages, we can share deeply the experiences of other humans as well as collect this information in forms that can be passed from one generation to another in ever-increasing sophistication. Without the capability that languages provide to transmit complex thought patterns, the amazingly complex, if not wonderful, civilizations in which the temples of Greece could be designed and built, in which Bach and Beethoven could compose, or in which Newton, Darwin, and Einstein could think would never have developed.

As extraordinary as we are, however, there is no reason to assume that we will necessarily continue to exist. No guarantee exists that the human species shall not be eliminated by some virulent disease entity or through the action of man-made toxic products, in particular the radioactive compounds that would be unleashed as a result of an unlimited nuclear exchange. Thus man, like all other forms of life, has no guaranteed "right" of perpetual existence. Only if as a species we continue to behave sensibly will we have a realistic chance to go on. Our survival, at least within the lifetime of the earth's current solar environment, thus always must be seen as in our own hands.

Our rights as humans thus come to us not from any inherent order (purpose) within the universe but as reflections of religious and legal edicts that we have acquired during the course of human history. The Ten

Commandments, for example, express sensible rules imposed to let us work together effectively toward the common goals that have made our Western civilization so successful. Not surprisingly, human rights have varied from one civilization to another, and while each of us as individuals may have very strong ideas as to what types of freedom are generally needed for meaningful lives, those freedoms we wish the most vary from time to time, depending upon the respective environments in which we find ourselves.

Just as it has been useful to convey specific rights to ourselves, we have also conveyed them to selected animals like the dog, cat, or cow that have long coexisted with humans and whose absence would bring us sorrow or economic distress if they were to be arbitrarily eliminated by other humans. Until recently, however, virtually no one regarded all forms of even higher life (vertebrates) as inherently worthy of protection. Pests were pests that should be removed if at all possible from interactions with human civilization. Rattlesnakes, coyotes, and rats were invariably perceived as animals to be eliminated, not protected. Today, however, ironically, most likely because of a gradual subconscious acceptance of the theory of evolution, there is a growing tendency among many diverse groups of people to regard the human imposition of harm (be it death or merely pain) on any form of life to be an essentially wicked act. For such people, the killing of the fox is incarnate evil, even though this means more chickens will be eaten by foxes. And even the mice and rats that we used to call vermin now have very vocal protectors who are disturbed if we use these rodents to search for cancer-causing chemicals or for experiments to discover the molecular uniqueness of the various forms of cancer. To such people, mice and rats have inherent rights that must be taken into account when we experiment upon them. I, however, find this concept totally without any merit. If it is not effectively challenged, even its partial acceptance will impede further medical research and limit its potential for reducing human misery.

Unfortunately, the current mood among government officials is to compromise with the animal-rights fanatics, with much credence given to the use of "alternative forms of testing" that do not involve living animals. I fear that such temporizing capitulation will lead to even further capitulation. Instead, we should stand our ground and label nonsense for what it is. To do so, we must have the courage to say that all forms of life are not necessarily good per se and that if the quality of human life is to be improved, we must experiment upon animals such as mice. We should get on with our jobs without feeling that we are in any sense heartless because we lack compassion for the mouse. Our survival as humans demands that we put ourselves first, and to think for a long time in a different fashion would be to invite extinction. Of course, this is not to say that we condone any senseless diminution of the earth's still multitudinous forms of plants and animals. I, for example, object greatly to reducing the numbers of those most remarkable of animals, the whales.

Mankind's current inability to feel nonmystically about animal life manifests itself in its current unease with recombinant DNA experimentation. Just as most people still want to believe that there is something more extraordinary than molecules at the heart of living existence, they cannot easily accept that DNA is all that important, being the crucial difference between the various forms of life, without endowing it with virtual genie-like properties that we will never be totally able to handle. Many individuals thus fear that in subjecting the potential hidden genies to the abnormal manipulations of recombinant DNA, we may stir them into revengeful acts that would incalculably harm the human species. As a result of such worries, the mere process of using recombinant DNA is now being singled out as much more potentially dangerous than other breeding and selection processes, which, of necessity, also yield new gene combinations that have never existed before. Thus, a herbicide-resistant plant, if created through recombinant DNA procedures, must be subjected to painstaking safety checks, whereas no such checks are required if it were generated, say, through the recently developed somatoclonal selection procedures.[1] The ultimate irrationality of this situation is now more than clear not only to scientists but also to many giant industrial corporations who are attempting to create commercially useful new plants.

This preoccupation with DNA as a potentially uncontrollable genie explains, in part, the concern that many individuals have about current

---

[1]It is a feature of plant tissue, which can be grown in culture in the laboratory, that it often exhibits a high degree of genetic variability, resulting from rearrangement in the chromosomes. Plants can tolerate such disruptions much more readily than animals, and agronomists take advantage of this property to select for desirable genetic traits. This process does not involve genetic engineering.

attempts to develop procedures for curing genetic diseases using DNA therapy. Also behind much of this hesitation is the feeling of many people that the DNA within us is not the result of stochastic processes, but of design, and that we have no right to interfere with the destiny that lies behind our individual existences. To me, however, the moral question to be settled is whether such gene therapy procedures have reasonable chances of effecting real cures. It is hoped that this dilemma will evaporate once it becomes clear that gene therapy is a valid procedure for curing sick individuals.

Much more difficult to resolve will be the question of whether selective abortion of genetically disabled fetuses should occur. Here again, the heart of the problem lies in how we perceive human life per se. Does its very existence demand that we cherish and protect it? Those of us who are scientifically knowledgeable about DNA instinctively act on the premise that the random processes that govern the replication of DNA are bound to lead to some human fetuses that can never grow up into happy, functional individuals. Must these genetically damaged fetuses be allowed to develop into babies whose suffering is bound to bring unmitigated pain not only to themselves, but also to their parents and all others who must try to help them? Should we deny the existence of what we perceive to be the essence of a real human life, the capacity to develop into a person who by interacting successfully with others helps make this a more interesting and compassionate world?

With time, we can hope that those individuals who give us worldwide governmental and religious leadership will, through their respective paths of either reason or revelation, resolve the dilemma of whether a life without a realistic chance to develop into an effective person must go out into the daunting conditions that the earth will always provide.

Ethical Implications of the Human Genome Project

1994

T here is now much widespread concern about the moral consequences of the ever-increasing genetic knowledge resulting from
the Human Genome Project. Ongoing now for some six years, this
collaborative international effort aims to identify the some 100,000 genes
that provide the information underlying the development and subsequent
adult functioning of human beings. Because our genes are so crucial to our
potential for meaningful lives, the ability to look at their precise, individual forms will provide ever more important tools for predicting the future
course of given human lives. Clearly such knowledge can be used for bad
purposes as well as good, and there now exists justifiable concern that we
may be opening a Pandora's box. Will we unintentionally blight the lives of
all too many humans that do not want their innermost essences revealed
either to themselves or to others? This knowledge, in the wrong hands,
might easily make their genetically damaged lives even more difficult to
rise above.

Such threats of potential misuse, however, should always be compared
to present and future suffering due to genetic diseases, whose unhappy
prognostic courses we can now only marginally affect. To alleviate this current suffering and to help prevent many future victims of genetic injustice
from being created, we need to find and study the genes that give rise to
these diseases. Once we have found the genetic culprits, we can rationally
search for pharmaceutical cures as well as to try to develop curative genetic therapies. And for those individuals who want to use them, we can provide diagnostic antenatal screening techniques to tell prospective parents

whether a fetus carries a gene whose possession will strongly impair its potential for meaningful life. Thus, instead of solely focusing our ethical concerns on potential misuses of this new knowledge, we must also face up to the moral consequences of not acquiring and using this new genetic knowledge. Not moving ahead to prevent new cases of genetic diseases or to search for cures, when we have the means to do so, also has strong ethical implications. Need a woman who already has one son with genetically impaired intelligence have to give birth to another son with the same tragic impairment?

Until less than two decades ago, we did not have the power to so alter our genetic destinies. Except for Down's syndrome, which already by 1959 was known to be caused by an extra copy of chromosome 21, we did not have the means to do prenatal diagnosis of any genetic disease. Then with the 1973 development of the powerful recombinant DNA procedures, it became possible to isolate and study specific fragments of DNA. Human DNA could for the first time be directly looked at, revealing millions of single base differences between any two individuals. Such polymorphisms were soon cleverly used to make the first genetic maps of humans, the reference points being the location of the polymorphisms along the 24 different human chromosomes. With this advance, human genetics could at last develop into a rigorous intellectual discipline. Until this big step forward, the obvious inability of human geneticists to do genetic crosses prevented us from knowing the chromosome location of the genes underlying most known genetic diseases.

Soon the dreaded neurological affliction, Huntington's disease, was located on chromosome 4, with cystic fibrosis (CF) soon after being found on chromosome 7. Knowledge of the chromosome site of a disease gene map, however, is but the first step in what often proves to be a very time-consuming goal of the isolation (cloning) of a disease gene. This has become the main immediate objective for most scientists interested in human disease. Once the gene, as opposed to chromosomal location, is found, genetic diagnosis becomes much more straightforward. Moreover, from the DNA sequence we can infer the amino acid sequence of the respective protein product which often provides vital information as to the function of the protein in cells. The cloning in 1989 of the CF gene, for example, immediately reinforced previous speculations that its protein product was a membrane protein involved in chloride transport, immediately suggesting routes for possible drug therapy. And with this gene now in hand, the possibility also exists for gene therapy research where we add copies of the good gene to diseased cells, with the hope that the respective good genes will function and cure the disease. Thus, several major attempts are now being made to introduce good copies of the CF gene into lung epithelial cells of CF patients.

Always making the cloning of any human disease gene a major research challenge is the very large size of the human genome, which contains approximately three billion (3×10^9) base pairs distributed over the 24 human chromosomes. Even when a disease gene is common enough to be mapped close to a well-defined DNA polymorphism, the DNA fragment corresponding to this map region might contain 5–10 million base pairs. Obviously, the better our genetic maps, the easier it should be to find a given disease gene, with the ultimate resource needed being the complete DNA sequence of the human genome. With it at our disposal, we could automatically scan a large fragment of DNA to see how many different genes we can spot, always looking to see if specific base-pair changes invariably are found associated with the disease we are trying to understand.

But when the task of how to sequence the entire human genome was first discussed in 1984, there were strong feelings that the time for this project had not yet arrived. The largest DNA molecule then sequenced contained less than 10^5 base pairs. No one individual, moreover, had single-handedly sequenced more than 20,000 base pairs within a given year. Without the development of much new technology, the complete sequencing of the human genome would require thousands of scientists working 10–20 years.

On the other hand, until the decision was made to go for the whole human genome, the majority of families suffering from genetic diseases would have no reason to hope for release from the pains inflicted upon them by past errors in the copying of their family's genetic messages. The first prominent scientist to go very public in support of the Human Genome Project was Renato Dulbecco, whom I have known since we first came together, more than 45 years ago, in Salvador Luria's lab at Indiana University. In the fall of 1985, Renato, in a speech at the Cold Spring Harbor Laboratory, spoke of the great potential advantages for cancer research that would result from sequencing the entire human genome, soon afterwards presenting his reasoning in *Science* magazine. My initial reaction, however, was of surprise, wondering why my highly intelligent friend believed that the time had come for such a massive endeavor.

Over the next several months, however, my initial misgivings had vanished, and I wanted to start the project as soon as possible. By then I saw it having two major purposes. Clearly, the first and the easiest reason to sell to the public would be its ability to vastly speed up the rate at which important disease-causing genes can be found. Here it's important to realize that most genetically predisposed diseases are not rare conditions limited to small numbers of families. Indeed they encompass a wide spectrum of common diseases like diabetes, arteriosclerosis, Alzheimer's disease, and many cancers. In the past, they have not been included among genetic diseases since the time of onset and severity has strong environmental components. For example, whether an individual comes down with late-onset diabetes is a function not only of his or her genetic constitution, but also of dietary history. Likewise, a combination of genes and diet determines an individual's probability of having a heart attack. And whether a woman comes down with breast cancer depends not only on the genes she inherits from her parents, but also upon her later exposure to cancer-causing agents that may be present in her food and in her environment.

Equally important, the human genome is our ultimate blueprint providing the instructions for the normal development and functioning of the human body. That we are human beings and not chimpanzees is not due in any sense to our nurturing but to our natures, that is, our genes. This does not mean that we have highly different sets of genes. In fact, both sets of some 100,000 genes carry out roughly the same biochemical tasks. But the five million years of evolution that has separated us from the chimps has led to significant divergencies from chimps as to the exact times at which some human genes function, as well as the rates at which they produce their respective protein products. Successive changes in certain key genes have led to the retention of many juvenile chimp features into adult human life. For example, the shape of our adult brain is that of the baby chimp brain, with our general lack of body hair resembling the situation of baby apes. As the Human Genome Project moves to completion, we will have the power to understand the essential genetic features that make us human. A key, though obviously not immediately ascertainable, objective is the set of instructions that lead to the development of those human brain features which give us the capacity for our written and spoken languages.

Nevertheless, there was much initial opposition to the Human Genome Program, certain aspects of which still persist today. Initially the main opposition came from within the scientific community, in particular from the molecular biologists who feared that the Human Genome Project would consume so much money that more immediately important research objectives would be threatened. By now this opposition has almost vanished, partly through the realization that the three billion dollars to be spent over 15 years for the Human Genome Project would consume less than two percent of the worldwide budget for biomedical research over that period. Equally important, by extending the scope of the Human Genome Project to include smaller genomes of model organisms like *E. coli*; the budding yeast, *Saccharomyces cerevisiae*; the roundworm, *C. elegans*; and the fruit fly, *Drosophila melanogaster*, a wide spectrum of the biological community saw immediate benefits for their own forms of fundamental research.

The second form of opposition arose from fears as to the ethical, legal, and social consequences of precise human genetic information. In particular, worries were expressed that we would be generating new forms of discrimination in which individuals, genetically predisposed to serious diseases, would be unable to obtain or hold jobs as well as obtain life and health insurance. The question of how to protect genetic privacy was clearly central to these concerns. Moreover, there were anxieties as to whether we as human beings would be able to cope psychologically with the ever-increasing knowledge about our genes that will soon be at our disposal. And how soon will it be before we are able to give our citizens the genetic education needed for them to make informed choices as to what information they want and then the consequences of using or not using these facts? To try to begin to address these problems, when the Human Genome Project was initiated within the United States, we made the decision to create a specific new program to consider the ethical, societal, and legal issues (ELSI) of the Genome Project. At its start in 1989, we devoted 3% of our genome monies to such issues, with now 5% of such funds going to these ELSI programs.

To advise NIH's Center for Human Genome Research on how these ELSI funds should be spent, a program advisory committee was set up drawn largely from users as opposed to generators of human genetic information. Heading this committee was the psychologist-turned-human geneticist, Nancy Wexler, at risk herself to Huntington's disease from which her mother had died. The remaining members represented a wide spectrum of ethnic backgrounds and political viewpoints who accepted their appointments with the belief that their actions would hopefully lead to a fairer and more humane introduction of genetic data to the American public. They, however, never saw their role to be the making of ethical decisions for the United States as a whole. No mandate existed for such a role, with such responsibility belonging to the executive and/or legislative branches of our government.

Complicating the ELSI agenda has been the increasingly obvious fact that different genetic diseases present quite different spectra of ethical, legal, and social dilemmas. Handling the complexities arising from cystic fibrosis, whose victims now can live into their thirties if not longer, is bound to be different from our approaches to diseases like Tay-Sachs, whose victims live in almost perpetual agony during their brief 1–2 years of life. So, not surprisingly, the now five years of ELSI programs and recommendations have not led either to new administration-based regulations or to broad-ranging legislative mandates. Only the State of California has passed any law aiming to ban genetic-based dissemination by health and life insurance companies. And nobody has specifically come up with regulations or laws protecting the privacy of individual genetic records. When looked at closely, however, neither genetic privacy nor genetic-based insurance dilemmas have simple, straightforward solutions. Thus, in the absence of cases of bad abuse, we should be careful not to rush into legislative and administrative actions whose long-term consequences cannot seriously be predicted.

Even when appropriate satisfactory laws and regulations are in place, there will still be many dilemmas that cannot easily be handled by these means. What responsibility, for example, do individuals have to learn about their genetic makeups prior to their parenting of children? In the future will we be generally regarded as morally neglectful when we knowingly permit the birth of children with severe genetic defects? And do such victims later have legal recourse against their parents who have taken no action to help prevent their coming into the world with few opportunities of living a life without pain and emotional suffering?

Those who think this way have to face the counterarguments that if we try to control the genetic destiny of our children, we are in effect eugenicists following in the dreadful footsteps of the Nazis, who used genetic arguments when they used gas chambers to kill some 250,000 already sterilized inhabitants of their pre-World War II mental institutions. While early in the twentieth century eugenics had almost the ring of a prospective movement attracting the support of many prominent Americans, the horrors of these later Nazi aspects has to make us fear that eugenic arguments might again in the future be used to promote the elimination of supposed genetically unfit political philosophies or ethics groups. Such worries come not only from minority groups who fear for their futures, but also, in particular, from many Germans whose revulsion of their own history leads them to oppose any aspect of genetic engineering, believing the technologies so developed provide a slippery slope for rekindling of Nazi-type eugenics actions.

Equally strong opposition to programs aimed at preventing the birth of severely genetically impaired children comes from individuals who believe that all human life is a reflection of God's existence and should be cherished and supported with all the resources at human disposal. Such individuals believe that genetically impaired fetuses have as much right to exist as those destined for healthy, productive lives. But such arguments present no validity to those of us who see no evidence for the sanctity (holiness) of life, believing instead that human as well as all other forms of life are the products not of God's hand but of an evolutionary process operating under the Darwinian principles of natural selection. This is not to say that humans do not have rights. They do, but they have not come from God but instead from social contracts among humans who realize that human societies must operate under rules that allow for stability and predictability in day-to-day existence.

Foremost among these rules is the strict prohibition in virtually all societies against the killing of a fellow human being unless necessary self-defense is involved. Without this rule, our lives as functioning humans would be greatly diminished with no one able to count on the continued availability of those we love and depend upon. In contrast, the termination of a genetically damaged fetus should not diminish the future lives of those individuals into whose world it would otherwise enter. In fact, the prevailing emotion must largely be one of relief of not being called to give love and support to an infant who never can have an existence whose eventual successes you can anticipate and share.

Thus, I can see only unnecessary agony from laws that use the force of arbitrary religious revelations to impose the birth of genetically sick infants upon parents who would much prefer to terminate such pregnancies, hoping that their next conception leads to a healthy infant. Using the name of God to let unnecessary personal tragedies occur is bound to upset not only those who follow less dogmatic guidelines for life but also many members of those religious groups whose leaders proclaim the absolute sanctity of all human life. These latter people are bound to ask themselves whether the words of God, as then interpreted, are more important than the health of their children or those of their friends. In the long term, it is inevitable that those authorities who ask their followers to harm themselves in the name of God will increasingly find themselves isolated with their moral pronouncements regarded as hollow and to be ignored.

Nonetheless, we would not be surprised by increasing opposition to the Human Genome Program, which is seen as the most visible symbol of the evolutionary biology/genetics-based approach to human existence. But because the medical objectives of the Genome Project cannot easily be faulted, those who fear its implications will emphasize that its reductionist approach to human existence fails to acknowledge the overriding importance of the spiritual aspect of human existence, which they will argue is much more important than our genes in determining whether we are successes or failures in our lives. Under that argument, we would be making better use of our monies in trying to improve the economic and moral environments of humans as opposed to the finding of genes that they believe will only marginally affect our health and social behavior.

With time, however, the truth must emerge that monies so spent have effectively no chance of rolling back the fundamental tragedies that come from genetic disease. So I believe that over the next several decades we shall witness an ever-growing consensus that humans have the right to terminate the lives of genetically unhealthy fetuses. But there remains the question as to who should make the decisions that lead to the termination of a pregnancy. Under no circumstances should these choices be assigned to the State, because even in our more homogeneous cultures, there exists wide divergency as to what form of future human life we have to encourage. Instead such decisions are best left solely in the hands of the prospective mother and father (if he effectively plays this parental role).

Such unregulated freedom clearly opens up possibilities for irresponsible genetic choices that only can harm all concerned. But we should not expect perfect results in handling genetic dilemmas any more than we can expect them from other aspects of human life. And we will have reason to hope that our genetic choices will improve as general knowledge of the consequences of bad throws of the genetic dice become better appreciated. Clearly we must see that genetics assumes much more prominent places in our educational curricula. Equally important, the appropriate genetic screening procedures must become widely available to all our citizens regardless of their economic and social status.

At the same time, we must always be aware that the human society will only come to our genetics way of thinking haltingly. Even many of our firmest supporters will worry at times that we may be moving too fast in assuming the roles that in the past we have assigned to the gods. Only they could predict the future as well as have the power to change our future fates from bad to good or from good to bad. Thus today we have some of those same powers. Clearly this is a situation that is bound to make many people apprehensive, fearing we will misuse our powers by helping create immobile, genetically stratified societies that do not offer the prospect of hope and dignity for all their citizens. Thus, in so moving through genetics to what we hope will be better times for human life, we must proceed with caution and much humility.

Genes and Politics

1997

The science of genetics arose to study the transmission of physical characteristics from parents to their offspring. When closely studied, much variation exists for virtually any characteristic, say, in size or color, among the members of all species, be they flies, dogs, or ourselves, the members of the *Homo sapiens* species. The origin of this variability long fascinated the scientific world, which already in the nineteenth century asked how much of this variation is due to environmental causes (nurture) as opposed to innate hereditary factors (nature) that pass unchanged from parents to offspring. That such innate heredity exists could never be realistically debated. One need just look at how characteristics in the shape of the face pass through families. Ascribing, say, the uniqueness of the Windsor face to nurture as opposed to nature goes beyond the realm of credibility.

Genes as the Source of Hereditary Variation Both within and between Species

The key conceptual breakthrough in understanding the nature component of variation came in the mid-1860s from the experiments of the Austrian monk and plant breeder, Gregor Mendel (1822–1884). In his monastery gardens he created, by self-breeding, strains of peas that bred true for a given character like pea color or pod shape. Then he crossed his inbred strains with each other and observed how the various traits assorted in the progeny pea plants. In his seminal scientific paper, published in 1865, Mendel showed that the origin of this hereditary variability lay in differences in discrete factors (genes) that pass unchanged from one plant generation to another.

Most importantly, he showed that each pea has two sets of these factors, one coming from the male parent, the other from the female. Some
of those factors are expressed when present in only one copy (dominant
genes), whereas others become expressed only when two copies, one from
each parent, are present (recessive genes). Mendel's results later were used
by the Danish botanist, Wilhelm Johannsen (1857–1927), to make the
important distinction between the physical appearance of an individual
(its phenotype) and its genetic composition (genotype). Mere examina-
tion of a plant's physical appearance need not reveal its genetic composition. Recessive genes present in only one copy can be identified only by
further genetic crosses. Mendel further made the equally important observation that genes do not necessarily stay together when the male and
female sex cells are formed. Instead, they often independently assort from
each other, giving rise to progeny with sets of features very different from
those of either parent.

Mendel's work, done before the behavior of chromosomes during cell
division was understood, almost had to lay unappreciated until the turn of
the century, when three plant breeders working on the European continent—Correns, De Vries, and Tschermak—independently rediscovered
the basic rules for hereditary transmission, which today we call Mendel's
laws. It was not until 1890 that the sex cells were found to possess only half
the number of chromosomes present in adult cells. Fertilization through
combining the haploid N number of chromosomes of the sperm with the
haploid N number of the egg restores the $2N$ diploid chromosome number of adult plants and animals. Except for those special chromosomes
that determine sex, adult cells contain two copies of each distinct chromosome, each of which is exactly duplicated prior to the cell division. With
the basic facts of chromosome behavior so established for both ordinary
cell division (mitosis) and sex cell formation (meiosis), the rediscovered
laws of Mendel were given a chromosomal basis by the American, Walter
Sutton. Perceptively, he noted in 1903 that the segregation patterns of
Mendel's genes exactly parallel the behavior of chromosomes during the
meiotic cell divisions that produce the male and female sex cells (the Chromosomal Theory of Heredity). During the next several decades, an ever-
increasing number of genes were found to have precise locations along
specific chromosomes. In essence, each chromosome came to be seen as a
linear collection of genes running between its two ends.

Genes first were of interest because they were the source of the variability between the members of a species, but they soon began to be appreciated more properly as the source of information that gives an organism its unique form and function. Its collection of genes (its genome) is what gives each organism its own unique developmental pathway. A dog is a dog, a bacterium a bacterium, and so on, because of the information carried by their respective genomes. Thus, gene duplication prior to cell division must be based on a very accurate copying process. Otherwise, there would be no constancy of genetic information and of the development processes they make possible. Correspondingly, genetic variation arises when genes are not accurately copied (mutated) and give rise to changed (mutant) genes.

Hereditary Variability Generated by Changes in Genes (Mutation) Underlies Evolution by Natural Selection

As soon as the first spontaneous gene mutations became known, they were perceived as the obvious source of the new genetic variants necessary for Darwinian evolution by "survival of the fittest." Many more dysfunctional than more functional genes, however, resulted from random mistakes in the gene-copying process. Thus, the rate at which the gene-copying process makes mistakes is likely also to be under strong evolutionary pressure. If too many spontaneous mutations occur, none of the mutant-gene-bearing organisms are likely to develop and produce viable offspring. Correspondingly, too low a mutation rate will not generate sufficient gene variants to allow species to compete effectively with those species evolving faster because of their more frequent generation of biologically fitter offspring.

Eugenic Solutions for Human Betterment

The coming together of Darwinian and Mendelian thinking immediately raised the question of the applicability of the new science of genetics to human life. To what extent was human success due to the presence in their recipients of good genes that led to useful biological traits like good health, social dependability, and high intelligence? Correspondingly, how many individuals at the bottom of the human success totem pole were there because they possessed gene variants perhaps useful for earlier stages in human evolution but now inadequate for modern urbanized life? Social Darwinian reasoning viewed the sociocultural advances marking humans' ascent from the apes as the result of continual intergroup and interpersonal strife, with such competitive situations invariably selecting for the survival of humans of ever-increasing capabilities. Social Darwinism came naturally to the monied products of the industrial revolution, a most prominent one being the talented statistician, Francis Galton (1822–1911). Early in his career, he wrote the 1869 treatise *Hereditary Genius*, later coining the term "eugenics" (from the Greek meaning wellborn) for studies that would bring about improvements of the human race through the careful selection of parents.

Clever though he was, and able to take comfort that he was Charles Darwin's (1809–1882) cousin, Galton's eugenic prescriptions offered no basic improvement on the long-attempted practice of seeing that offspring from families of attainment married into families of similar high function. In this way, supposed good germ plasm would not be diluted by inputs of putative bad heredity. But whether Galton was promoting reality, as opposed to an unjustified prejudice against the vulgarity of the lower classes, had no way of being even half-tested before the arrival of Mendelian analysis. So the eugenics movement naturally became galvanized by the new laws of Mendelian heredity. But immediately, their hopes had to be tempered by the fact that human genetics never would have the power of other forms of genetics where genetic crosses could be made as well as observed. For better or worse, the eugenicists' main research tool had to be hopefully well-collected, multigenerational pedigrees of physical and mental traits that passed through families from one generation to the next. Toward that end, Galton, then already 84, co-authored in 1906 the book *Noteworthy Families*, an index to kinships in near degrees between persons whose achievements are honorable and have been publicly recorded.

Initially, there were hopes that simple Mendelian ratios would characterize the inheritance of a broad-ranging group of human traits. But in addition to the limitations brought about through the inability to confirm genetic hypotheses through genetic crosses, many of the studied traits appeared in too few families for appropriate statistical analysis. Particularly difficult to analyze were progeny traits not present in either parent. Conceivably, individuals had inherited one copy of the same recessive gene from each parent. Such tentative conclusions became more convincing when the respective traits, like albinism, were found more often in highly inbred, isolated populations where marriages of cousins were frequent.

Easier to assign as bona fide genetic determinants were dominant-acting genes that need be inherited from only one parent for their presence to be felt. Once Mendelian thinking had appeared, the inheritance mode of Huntington's disease, the terrible neurological disease that leads to movement and cognition disorders, was quickly ascertained as a dominant gene disorder. Similar clear genetic attributions could be assigned to traits, such as red–green color blindness and hemophilia, which preferentially appear in males but which are never passed on to their own male offspring. This is the behavior of a trait caused by a gene present on the X sex chromosome, two copies of which are present in females but only one in males, whose sexuality is determined by the Y chromosome.

Important as these diseases of the body were to the individuals and families of those so afflicted, the main focus of early twentieth century eugenicists soon moved to potential genetic causations for disabilities of the mind, embracing a wide spectrum of manifestations from insanity through mental defectiveness, alcoholism, and criminality, to immorality. With poorhouses, orphanages, jails, and mental asylums all too long prominent features of the most civilized societies, eugenicists with virtually religious fervor wanted to prevent more such personal and societal tragedies in the future. They also desired to reduce the financial burdens incumbent on civilized society's need to take care of individuals unable to look after themselves. But in their evangelical assertions that genetic causations lay behind a wide variety of human mental dysfunctions, the early eugenically focused geneticists practiced sloppy, if not downright bad, science and increasingly worried their more rigorous geneticist colleagues.

American Eugenics: Sloppy Genetics for the Legitimation of Class Stratification

The most notable American eugenicist, whose conclusions went far beyond his facts, was Charles B. Davenport (1866–1944), who parlayed his position as Director of the Genetics Laboratory at Cold Spring Harbor, New York, to establish in 1910 a Eugenics Record Office using monies provided by the widow of the railway magnate E.H. Harriman. In his 1911 book *Heredity in Relation to Eugenics*, pedigrees were illustrated for a wide-ranging group of putative hereditary afflictions ranging from bona fide genetic diseases, such as Huntington's disease and hemophilia, to behavioral traits of much less certain hereditary attribution, such as artistic ability and mechanical ability with reference to shipbuilding. With so little then known about the functioning of the human brain, Davenport's early rush to associate highly specific accomplishments of the human brain to specific genetic determinants could not automatically be dismissed as nonsense. In today's intellectual climate, however, a predilection for genes that predispose individuals to city life as opposed to rural life would not be the way to an academic career. But, even his fellow early eugenicists must have regarded as more wartime patriotism than science his 1917 claim that a dominant gene for thalassophilia predisposed its recipients to careers as naval captains.

In addition to its family pedigree assembly and archival roles, the Cold Spring Harbor Eugenics Record Office frequently counseled individuals with family backgrounds of genetic diseases, particularly when they were considering marriage to blood relatives. Many such seekers of help must have been misled by advice that never should have been given, considering that era's limited power for meaningful genetic analysis. Worries about insanity were a major concern, where manic–depressive disease was seen to move through some families as if it were a dominant trait. In contrast, schizophrenia had more aspects of a recessive disease. Yet, even with today's much more powerful human genetics methodologies, we still do not know the relative contribution of dominant versus recessive genes to these two major psychoses or any other form of mental disease.

The Eugenics Record Office's pre-World War II message was that insanity usually expressed itself only when genes predisposing it were inherited from both parents. If this were so, siblings of individuals displaying mental instability were at risk of being carriers of insanity-provoking genes. Because recessive genes for insanity would be silently passing through many families, marriage to any individual with mentally disturbed siblings was not prudent. Even more certainly, marriage should be avoided between individuals having severe mental illness in both parents. In those days, when no effective medicines existed for any form of psychiatric illness, most families bearing mental disease not surprisingly kept this knowledge as secret as possible. There must have been many couples, perhaps overworried about producing mentally disturbed offspring, who chose not to have children.

The eugenicists predictably were concerned about mentally unstable individuals marrying those with similar disturbances. Also of obsessive concern to them were individuals with feeble-mindedness, where Davenport believed that recessive genes were also involved. With his certainty that all children of two feeble-minded parents would be defective, he wrote of the "Folly, yes the crime, of letting two such persons marry." In his mind, the inhabitants of rural poorhouses were there largely because of their feeble-mindedness, and he considered one of our nation's worst dangers to be the constant generation of feeble-minded individuals by the unrestrained lusts of parents of similar conditions.

It was to stop such further contaminations of the American germ plasm that Davenport, as early as 1911, saw the need for state control of the propagation of the mentally unstable or defective. Initially, he did not favor adoption of state laws allowing for their compulsory sterilization, an idea then considered wise and humane by much of that era's socially progressive elite. Clearly somewhat sexually repressed (obsessed?), he feared that with pregnancy no longer a worry, the sexual urges of the sterilized, mentally unstable impaired might cause more harm to society than even the procreation of more of their kind. Instead, he wanted mentally impaired women to be effectively segregated (imprisoned?) from the impaired of the opposite sex until after they passed the age of procreation. This prescription, however, was totally unrealistic and the American eugenics movement as a whole enthusiastically promoted the compulsory sterilization legislation that spread to 30 states by the start of World War II.

If the eugenics movement had focused its attention predominately on genetic afflictions that truly disabled its recipients, we might now be able to look back at it as a mixture of sloppy science and well-intentioned but kooky naiveté. Photos of the eugenics booths of the 1920s state farm fairs are virtually laughable. In them can be seen "fitter families" displayed near the pens at which prize cattle were shown. The thought that sights of their earnest faces would lead to preferential procreation of more of the same now stretches our credulity. In contrast, the words and actions of Harry P. Laughlin, Davenport's close associate and Superintendent of the Eugenics Record Office, today can only make our minds flinch.

Pleased that his ancestors were traceable to the American Revolution, Laughlin shared Davenport's belief that the strengths and weaknesses of national and religious groups were rooted in genetic as well as in cultural origins. While, at least in public, Davenport wrote that no individual should be refused admission to the United States on the basis of religious group or national origin, Laughlin stated as scientific fact before appropriate Congressional bodies that the new Americans from Eastern and Southern Europe were marked by unacceptable amounts of insanity, mental deficiency, and criminality. Although he lacked any solid evidence, he nonetheless promoted the belief that the newest immigrants to our shores were much more likely to be found in prisons and insane asylums than were the descendants of earlier waves of English, Irish, German, and Scandinavian settlers. Even though the then-current postwar hysteria against unrestrained immigration by itself might have led to the 1924 legislation, there is no doubt that Laughlin's testimony tilted the composition of the future immigrants to Northern Europeans.

With legislation in place, Davenport no longer had to fear that "the population of the United States will on account of the great influence of blood (genes) from South-Eastern Europe rapidly become darker in pigmentation, smaller in stature, more mercurial, more attached to music and art, given to crimes of larceny, kidnapping, assault, murder, rape, and sex immorality and less given to burglary, drunkenness, and vagrancy than were the original English settlers." Through propagating such racial and religious prejudices as scientific truths, the American eugenics movement was, in effect, an important ally of the ruling classes, many of whose privileges inevitably came through treating those less fortunate as inherently unequal.

Using the First IQ Tests to Justify Racial Discrimination within the United States

The emergence of intelligence measuring reinforced the belief of America's prosperous people that their wealth reflected their respective family's innate intellectual superiority. The French psychologist, Alfred Binet (1857–1911), was the first person to try to systematically measure intelligence, responding to a 1904 request from the French government to detect mentally deficient children. The resulting Binet–Simon tests crossed the Atlantic by 1908, being first deployed in the United States by Henry Goddard in New Jersey at a training school for feeble-minded boys and girls. Soon afterward, he went on to test 2000 children with a broad range of mental abilities. Initially, there was considerable public opposition to the testing of "normal" individuals because of the test's first use on the feeble-minded. Within only a few years, however, revised Binet–Simon tests, more appropriate for precocious children, were prepared by Lewis Ternan (1877–1956) at Stanford University. These so-called IQ (intelligence quotient) tests were soon employed during World War I on hundreds of thousands of army draftees. Their main function was not to weed out mental defectives, but to assign recruits to appropriate army roles. Those administering the tests, led by the noted psychologist Robert M. Yerkes (1876–1956), claimed they were seeing native intelligence independent of the recruit's environmental history. Yet, clearly, many of the questions or arithmetic problems would be more easily answered by those with extensive schooling and possessing a broad vocabulary. Not surprisingly, the non-English-speaking recruits just off the immigration boats tested badly, allowing a test leader to privately confide to Davenport, "we are well on the right track in our contention that the germ plasm (now) coming into the country does not carry the possibilities of that arriving earlier." Such "objective test data" further convinced the eugenicist world that not only was mental deficiency genetically determined, but so was general intelligence.

Although black men from urban areas tested higher than white southern rural men, their IQ scores were significantly lower than their white equivalents from the same communities. Given today's realization that intelligence measurements virtually by necessity have cultural biases, the comparative data assembled from the army recruits had little real meaning. In many ways, it was like comparing oranges with apples. Nonetheless, the data summarized in *Psychological Examining in the United States Army* were used to justify the discriminatory segregation laws that effectively made America's black population second-class citizens. Genetic inequalities across so-called race boundaries were taken for granted, and 29 states maintained laws against black–white intermarriages, often using the argument that the superior white germ stock would be diluted with inferior genes.

Although eugenics had its origin in England, it never affected the national consciousness there as it did in the United States. With social class stratification so long a characteristic feature of British life, the ruling classes had no need of further justification for their privileged existence. To a lesser but real extent, social inequalities also were taken-for-granted features of most European countries, many of which still had royal families and their attendant aristocracy. Enthusiastic prewar eugenics movements nonetheless sprang up all over the continent, extending even to Southern America and Japan in the 1920s. Everywhere, the chief adherents were the professional middle class, naively proselytized into believing that genetic thinking could soon lead to human beings with heightened hereditary capabilities. Although the continent's eugenicists frequently used the now unacceptable term "race hygiene" for their movement, their ways for the betterment of human heredity for the most part in no way infringed upon preexisting human liberties. Offered as the future panacea was the standard package of marriage between genetically healthy individuals, with correspondingly strong disapproval of marriage for individuals bearing obviously bad genes like those leading to Huntington's disease. Only in two European countries, Germany and Sweden, was legislation enacted for obligatory sterilization of individuals thought to be the bearers of disabling genes.

Nazi Eugenics (Race Hygiene): A Murderous Ménage á Trois of Bad Genetics, Racial Anthropology, and Psychiatry at the Beckon of Hitler and Himmler

Although it was the Hitler-led Nazi government that quickly passed the 1933 Eugenic Sterilization Law, the broadly based German eugenics movement of the 1920s laid the groundwork. Then it was embraced by a spectrum of political thought, much of it totally respectable by the ethical standards of those days. The Germany of that time was a nation undergoing a great moral crisis brought on by its humiliating defeat in the World War. Its four awful years of trench warfare had killed a significant fraction of its better younger men and left it vulnerable to the hyperinflation that wiped out much of the savings of its professional middle class. Unlike England or France, Germany as a world power had only a fleeting existence, and the German people then saw the need to somehow reinvigorate themselves. The eugenicists' vision that human beings' futures lie in their genes struck a receptive chord in the immediate postwar German psyche. Even in the postwar chaos of its Weimar government, human genetics gained strong governmental support. Genetics quickly became a high-quality science in Germany, with Berlin becoming one of the world's leading centers for genetics. Study of supposed genetic differences between the so-called races was vigorously promoted, with it being accepted as fact that the commercial colonization of the world by major countries in Europe and the United States reflected the inherent superiority of the Nordic people's genes for intelligence and strength of moral purpose. Anthropological-based research had strong genetic components, with genes being perceived as the crucial element determining human behavior.

In total contrast, genetic explanations for human successes were not favorably received in the Soviet Union, whose communist doctrines emphasized social, as opposed to genetic, causation for the currently existing inequalities between humans. Already by the mid-1930s, eugenic thinking had become strongly inimical to Russian Communist policymakers, who increasingly favored the Lamarckian explanations (inheritance of acquired characteristics) of Trofim Lysenko (1898–1982), its homegrown agriculturist, over the foreign-originating Morgan–Mendelian analysis of heredity. Those pursuing genes within the Soviet Union soon were putting not only their careers, but also their lives at risk. The great American geneticist, Hermann J. Muller (1890–1967), whose left-wing views led him to leave Texas and go to Russia in 1933, effectively ended his Soviet career when, in 1936, he compared Lamarckian thinking to alchemy, astrology, and shamism.

Seeking backing for the putative superiority of the Caucasian race, Adolf Hitler (1889–1945), while imprisoned in 1924 for the failed Munich putsch, read *Menschliche Erblichkeitslehre and Rassenhygiene* (The Principles of Human Heredity and Race Hygiene), a leading German genetic text of the time coauthored by E. Baur, Eugen Fischer, and Fritz Lenz. Enveloped by an uncritical eugenic perspective, it strongly reinforced Hitler's view of Germans as the *master race* that justifiably should rule the world. If, however, the Germans were indeed the master race, the Nazis had to explain their nation's humiliating defeat in the Great War and its subsequent devastating hyperinflation.

A perfidious scapegoat was needed, and here Hitler drew upon the long-existing, anti-Semitic feelings of many German people. Long segregated in rural enclaves dating back to the Middle Ages, Jews became effectively part of Germany's commercial and professional life only by the middle of the nineteenth century. Gravitating especially to the professions where their talents could more easily prevail over still-existing prejudices, Jewish importance in German commerce and professional life soon became disproportionate to their numbers, creating jealousies that inevitably fanned preexisting anti-Semitism. Clearly, much of this Jewish success reflected its religion and its respect for the intellect as opposed to oft revelatory-based opinions of their Christian equivalency. Their anti-Semitic opponents, however, saw the Jews' upward trajectory as manifestation of inherent immoralities that let them take unfair advantage of the more honest Christian Germans.

Until the arrival of Mendelian thinking, German anti-Semites never consistently decided whether their enemy was the Jews themselves or their religion. If their failure to acknowledge Christ was the problem, Jews who converted posed no further threat to Christian civilization. But if their reputed unscrupulous behavior and sexual licentiousness reflected innate hereditary qualities, their presence within Christian societies threatened their country's moral resolve, if not its very existence. Assertions by eugenicists that gene differences lay behind human behavioral differences were thus made to order for Nazi needs. From the 1933 start of their absolute rule, the Nazi propaganda machine ruthlessly portrayed Jews and Communists as the two main villains blocking the ultimate triumph of National Socialism. No words were vile enough to express their hatred for the genes that supposedly let Germany's one million Jews steal for themselves the monies and jobs of the honest Germans, or their horror of the Communists who wanted to redistribute monies from those who worked hard to those not able or willing to take care of themselves. Treated with equal contempt by the Nazis, but of less importance because of their much smaller numbers (30,000), were the Gypsies. Because of their wandering, supposed sexually unrepressed life styles, and their lack of respect for property, the German Gypsies were regarded by Nazi anthropologists as descendants of peoples of primitive etiological origin who had mated repeatedly with the German criminal, asocial subproletariat. So considered, the further breeding of this mixed-blood people must be stopped.

Sterilization of the Mentally Unfit as a Prelude to More Broadly Based Wartime Genocide

Gypsies, however, were not specifically targeted under the 1933 Eugenics Law that mandated compulsory sterilization for schizophrenia, manic–depressive psychoses, hereditary epilepsy, Huntington's chorea (disease), hereditary blindness, hereditary deafness, severe physical deformity, and severe alcoholism. Tribunals of Hereditary Health, consisting of a judge, a government medical officer, and an "independent" physician, made the resulting decisions where the individuals concerned often knew they were at risk only when called before its members. With appeals extremely difficult, these psychiatrist-led verdicts between 1934 and 1939 led to some 400,000 compulsory sterilizations, many to noninstitutionalized persons. These reputedly hereditary-damned individuals were further subjected to a subsequent 1934 law forbidding persons with serious mental disturbances from marrying. A year later, legislation specifically affecting Jewish marriages came through the 1935 "Nuremberg Decrees" for the protection of German blood and health. They forbade not only marriages, but also sexual intercourse, between the so-called German and Jewish races.

Concomitant with these eugenic actions was the assembly of vast record collections documenting individual hereditary–biological characteristics. To the Reich Kinship Bureau were referred decisions as to the origin of individuals with potential partial Jewish blood. Many such anthropological "expert conclusions" were made using only photos of the putative fathers. However, with the census of 17 May 1939 providing supposed "confidential" information of any Jewish grandparent, the Nazis, as the war started, felt they had a firm handle on the Jewish blood within their midst. So encouraged, later that year Professor Eugen Fischer, this time responding to coal barons of the Ruhr, wrote, "When a people wants somehow or other to preserve its own nature, it must reject alien racial elements and when they have already insinuated themselves, it must suppress and then eliminate them. The Jew is such an alien...."

With the war on, the German government, seeing no reason to waste scarce resources to keep what they considered genetically inferior peoples alive, proceeded to what it termed a "euthanasia policy of mercy killing." In a one-sentence letter postdated 1 September 1939, Hitler himself wrote, "Reichleiter Bouhler and Dr. Brandt are entrusted with responsibility of extending the rights of specifically designated physicians such that patients who are judged incurable after the most thorough review of their condition which is possible can be granted mercy killing." So authorized, 3000 mental patients in occupied Poland were summarily shot by storm troopers. In the Reich itself, where German citizens were involved, somewhat more formal procedures were used. Questionnaires were distributed to the mental hospitals, where they were completed in their capacity as experts by nine Professors of Psychiatry assisted by 39 other medical doctors. For their labors, they were paid 5 pfennigs (the cost of a cigarette) per questionnaire when they processed more than 3500 per month, but up to 10 pfennigs when fewer than 500 questionnaires per month were processed. The patients so selected for "euthanasia" had their respective questionnaires marked with a cross. Subsequently, carbon monoxide supplied by I.G. Farben was used for the elimination process. Before the killings stopped in the fall of 1941, some 94,000 mental patients had been killed. Subsequently, covert "euthanasia" by starvation, drugs, and failure to treat infectious diseases led to only 15% (40,000 persons) of Germany's prewar mental hospital population remaining alive at the war's end.

The primary reason for the supposed stopping of mental patient "mercy killings" was the need to transfer the personnel trained in killing by gas to the concentration camps, primarily in Poland (e.g., Auschwitz), to which most German Jews and gypsies had already been deported. With the decision already taken to invade the Soviet Union, a conference was held in March 1941 in Frankfurt at the Institute for the Investigation of the Jewish Question. At this conference, Dr. Gross, the head of the Race Policy Institute of the Nazi party, stated, "The definitive solution must involve the removal of Jews from Europe and he demands sterilization of quarter Jews." In an October letter to Himmler, Oberdiensleiter Brack of the Fuhrer's chancellery wrote that there are no objections to doing away (gassing) Jews who are unfit for concentration camp work. Less than a month later, Rosenberg, theoretician and minister of the occupied eastern territories, before representatives of the German press, announced the Final Solution of the Jewish Question, revealing plans, still to be kept secret, for the eventual mass murder of all European Jews, including the six million then living in the Soviet Union. The gas chambers so used were in no way restricted to Jews and Gypsies, with Soviet prisoners of war being victims of the first uses of Zyklon B (hydrocyanic acid) at Auschwitz. Anxious to give their racial extermination policies "scientific" justification, Himmler later in 1943 specified in a decree that only physicians trained in anthropology should carry out selection for killing and supervise the killings themselves in extermination camps. Some quarter Jews were to be spared, but not those with Jewish facial features who should be treated as half Jews. By the war's end, five to six million European Jews were so killed, the majority by the gassing procedures that the Nazis' co-opted human geneticists, psychiatrists, and anthropologists thought appropriate for individuals bearing genes inimical to the best interests of the German people.

With the liberation first of Poland and then Germany, the full horror of the racially based genocide policies of National Socialism quickly became known, generating even further disgust for the pseudoscientific theories of race superiority and purity that underpinned Nazi ideology. Anyone subsequently calling himself a eugenicist put his reputation as a decent moral human at risk. In fact, before the war even started, eugenics in the United States already was being perceived more as a social than a scientific movement. Already in 1930, the leaders of the Carnegie Institute of Washington had been told that its Cold Spring Harbor Eugenics Record Station practiced sloppy, if not dishonest, science. But with its founder Charles Davenport nearing retirement, it was allowed to expire more slowly than in retrospect it should have. Its doors closed only when Miloslav Demerec became director of the Department of Genetics in 1942. There thus was the embarrassment of Harry Laughlin's receipt in 1936 of an honorary degree from the University of Heidelberg in recognition of his contributions to racial hygiene. Undoubtedly pleased that eugenics, then fading in the United States, was becoming even more ascendant in Germany, Laughlin went to New York to receive his diploma from the German Diplomatic Counsel.

Eugenics, a Dirty Word, as the Search for the Chemical Nature of the Gene Begins

By the time I first came to Cold Spring Harbor for the summer of 1948, accompanying my Ph.D. supervisor Salvador Luria, then a professor at Indiana University, the Eugenics Record Office had been virtually expunged from its consciousness. Only in the library was its ugly past revealed through the German journals of the 1920s and 1930s on human genetics and race hygiene. No one that summer showed any interest in human genetics as a science or toward the general question of how much of human behavior reflects nature as opposed to nurture. Instead, genetic research there focused on the fundamental nature of genes and their functioning. It was not that human genetic diseases had suddenly become unimportant to its director, Miloslav Demerec. But there was general agreement both by the year-round staff and the many summer visitors that until the chemical identity of the gene was elucidated and the general pathways by which it controlled cell structure and functioning were known, it was premature to even speculate how genes contributed to human development and behavior.

Then, much sooner than anyone expected, the gene was revealed in 1953 to be DNA. The genetic code was established by 1966, and gene expression was seen to be controlled by DNA-binding regulatory proteins between 1967 and 1969. Genetics, happily then, had no reasons to intersect politics, except in Russia, where the absurdity of its Lamarckian philosophy became painfully more clear to its intelligentsia with every new major advance in molecular genetics. These major genetic breakthroughs were largely accomplished using the simple genetic systems provided by bacteria and their viruses that go under the name phages. By 1969, phage had become so well understood genetically that it became possible to create specific phage strains cleverly engineered to carry specific bacterial genes from one bacterial strain to another. Yet, seeming more ashamed than pleased with their neat science, James Shapiro and Jonathan Beckwith of Harvard Medical School held with much fanfare a press conference to announce that their new way to isolate specific genes was on the pathway to eugenically motivated genetic engineering of human beings. Knowing of the left-wing views of their "Science for the People" group, I, like most of my colleagues in the Boston region, saw their self-denunciations as manifestations of unrepentant leftist fears that further genetic research would render inviable the Communist dogma that assigned all social inequalities to capitalistic selfishness. Shapiro then moved (temporarily) to Cuba to regain his ideological purity.

Although the phage transductional system developed by Shapiro and Beckwith proved not to be a forerunner for eventual human genetic engineering, this was not true for the much more powerful and general "recombinant DNA" methodologies that Herbert Boyer and Stanley Cohen developed 4 years later, in 1973, just 20 years after the discovery of the double helix. Their new procedures allowed the isolation (cloning) of specific genes, through their insertion into tiny chromosomes (plasmids) that could be moved from one cell to another. At roughly the same time, unexpectedly powerful new ways to determine the exact sequences of the four letters (A, G,

T, and C) of genetic messages were worked out by Fred Sanger in Cambridge, England, and by Walter Gilbert and Alan Maxam at Harvard. Together, using these two new techniques, the exact structure of any gene could eventually be determined, given the appropriate facilities and resources.

The resulting recombinant DNA era, however, despite all the promises it held for major scientific advances, did not immediately take off. It initially stalled because of fears that among the many new forms of DNA created in the laboratory would be some that would pose unacceptable dangers to life as it now exists. In particular was the fear that highly pathogenic new forms of viruses and bacteria would be created. To give time to assess such potential dangers scientifically, a scientist-initiated moratorium on recombinant DNA research was declared in 1975. Effectively, it blocked virtually all recombinant DNA research for the next 2 years; and research concerned with cancer, where worries were expressed that a cancer gene might become bacterially transmitted, was held up for 2 more years.

During the moratorium, governmental committees were set up in the United States and in various European countries to assess the potential dangers from recombinant DNA experimentation in relation to its potential benefits for biology, medicine, and agriculture. No plausible scientific reasons for stopping such research emerged, and such committees, often containing public as well as scientific representatives, invariably concluded that in the absence of any quantifiable potential dangers, it would be irresponsible not to move ahead with experiments that could dramatically change the nature of biology. In retrospect, these decisions to move ahead were always the correct ones. For example, cancer research and our knowledge of the genetic basis of the immune system would effectively be back in the scientific middle ages if the enlightenments made possible through recombinant DNA had not occurred.

To my knowledge, moreover, not one case of recombinant DNA-induced illness has since occurred. No person has been so killed, nor has even one case of serious illness been attributed to recombinant DNA, nor do we know of any case where the release into nature of any recombinant DNA-modified organism has led to any known ecological disaster. This is not to say that someday a recombinant DNA-induced disease or ecological upset will not occur. Today, however, there is certainly no logical reason for not exploiting recombinant DNA procedures as fast as possible for human betterment.

Ideological and Value-based Oppositions to Recombinant DNA Research

Although there was no evidence of danger from recombinant DNA, there soon arose much visible and sometimes regretfully effective opposition to recombinant DNA research. Here the distinction should be made between objections from scientists who understand the technical issues involved and opposition from groups of public citizens who, though not understanding the science involved, nonetheless oppose much to all recombinant DNA research. Although some initial opposition arose from scientists whose own DNA research was not going well, virtually all the continuing scientific opponents at their heart had political hang-ups. As leftists, they did not want genes involved in human behavioral differences and feared that the onslaught of scientific advances that would follow from the unleashing of recombinant DNA might eventually allow genes affecting mental performance to be isolated and studied.

As a member of the Harvard Biology Faculty between 1975 and 1977, I watched in despair when "Science for the People" successfully assisted the public members of the Cambridge, Massachusetts, City Council to block recombinant DNA research at our Biological Laboratories. Later, I asked Salvador Luria, who was then at Massachusetts Institute of Technology and who knew that his left-wing friends were putting forth scientifically dishonest statements, why he never publicly criticized them. His reply was that politics was more important than science. This remark has long haunted me, because my own career owes much to the generous way he shared his great scientific talents with me at the beginning of my scientific career. But as a Jew who had to flee first his native Italy and then France for the eventual safety of the United States, Luria's left-wing political affinities were understandable, and I'm lucky I never had to so choose.

Specific political ideologies, however, are not the cause of the prolonged and sometimes effective opposition to recombinant DNA from parts of the general public, particularly in German-speaking regions. With professional agitators like Jeremy Rifkin playing important roles in heightening these public fears, such leadership would never have been effective if their audiences were at emotional ease with the gene and the geneticists who study it. The concept of genetic determinism is inherently unsettling to the human psyche, which likes to believe that it has some control over its fate. No one feels comfortable with the thought that we, as humans, virtually all contain one to several "bad" genes that are likely to limit our abilities to fully enjoy our lives. Nor do we necessarily take pleasure in the prospect that we will someday have gene therapy procedures that will let scientists enrich the genetic makeups of our descendants. Instead, there has to be genuine concern as to whether our children or their governments decide what genes are good for them.

Genetics as a discipline must thus strive to be the servant of the people, as opposed to these governments, working to mitigate the genetic inequalities arising from the random mutations that generate our genetic diseases. Never again must geneticists be seen as the servants of political and social masters who need demonstrations of purported genetic inequality to justify their discriminatory social policies. On the whole, I believe that genetics still commands broad respect in the United States and in much of Europe, despite the efforts of the recombinant DNA opponents to portray the genetic manipulations underlying the biotechnology industry as money-driven actions done at the expense of the public's health and the world's environment. Unfortunately, genetics and geneticists remain much less highly respected in Germany. There even today the most benign of recombinant DNA experiments remain controversial and subject to needless regulation. Propagation of genetically engineered plants is routinely sabotaged, with the mere practice of human genetics regarded as a criminal act by extremists on the left.

This German dislike for the gene and its human-directed manipulations is easily assignable to their Nazi eugenics past. The vile actions then done in the name of the gene hover as almost permanent nightmares never erasable from their national identity. As human beings, never sure that the world is immune from further such depraved behavior, we should never let this awful past slip from our consciousness. At the same time, the whole civilized world will suffer if today's German geneticists are unfairly thought to be cut from the same material that clothed those German geneticists, anthropologists, and psychiatrists who not only assisted the Nazi eugenic efforts, but promoted them as scientific-based necessities for German progress.

Part of today's problem may lie in the postwar fate of Hitler's biological conspirators. Naively as outsiders we long assumed that they would have all been treated as potential if not real war criminals, with even those of only slight guilt losing all further opportunities for academic existence. But as the German geneticist Benno Müller-Hill courageously pointed out in his 1984 book, *Todlicel Wissenschaft* (*Murderous Science*, Cold Spring Harbor Laboratory Press, 1998), there was no attempt by the German academic community to find out what truly happened. Instead, it was academically dangerous in Germany to explore the half-truths that allowed many key practitioners of Nazi eugenics to resume important academic posts. A number of professors who early joined the Nazi Party or SS and were directly involved with its genocide programs committed suicide, but there were many Nazi-assisting scientists, successfully claiming that they were only apolitical advisors, who slid quietly back into academic prominence.

The most damning example was that of Professor Otmar von Verschuer, who actively helped the Nazis—first at the Kaiser Wilhelm Institute of Anthropology under Professor Eugen Fischer and later at his own Institute of Human Genetics in Frankfurt. Involved in distinguishing Jews and part-Jews, he later closely collaborated with his former assistant, the now-notorious Joseph Mengele, then doing "scientific" research at Auschwitz. After the war, he nonetheless was appointed to be Professor of Human Genetics at the University of Munster. Equally disturbing was the postwar appointment of Fritz Lenz as head of an institute for the study of human heredity at the University of Gottingen, Germany's most distinguished university. Although clearly a very competent scientist, he was a major advisor for laws on euthanasia between 1939 and 1941, as well as author of a 1940 memorandum, "Remarks on resettlement from the point of view of guarding the race."

The postwar 1949 exoneration of von Verschuer occurred despite knowledge of the 1946 article in *Die New Zeit* accusing him of studying eyes and blood samples sent to him from Auschwitz by Joseph Mengele. Yet a committee of professors, including Professor Adolf Butenandt, later the head of the Max Planck Gesellschaft (the postwar name for the Kaiser Wilhelm Gesellschaft), concluded that von Verschuer, who possessed all the qualities appropriate for a scientific researcher and teacher of academic youth, should not be judged on a few isolated events of the past. I find it difficult to believe that the Butenandt committee had gone to the trouble of reading his article published in the *Volkischer Beobachter* 1-8-42. In it he wrote, "Never before in the course of history has the political significance of the Jewish question emerged so clearly as it does today. Its definitive solution as a global problem will be determined during the course of this war." Now there may be more reason to remember Professor Butenandt for his part in the von Verschuer whitewash than for his prewar Nobel Prize for research on the chemistry of the estrogen sex hormone.

Genuine Human Genetics Emerges from Recombinant DNA Methodologies

Long holding back the development of human genetics as a major science was the lack of a genetic map allowing human genes to be located along the chromosomes on which they reside. As long as conventional breeding procedures remained the only route to gene mapping, the precise molecular changes underpinning most human genetic diseases seemed foreordained to remain long mysterious. The key breakthrough opening a path around this seemingly insuperable obstacle came in the late 1970s when it was discovered that the exact sequence (order of the genetic letters A, G, T, and C) of a given gene varies from one person to another. Between any two individuals, roughly 1 in 1000 bases are different, with such variations most frequently occurring within the noncoding DNA regions not involved in specifying specific amino acids. Initially most useful were base differences (polymorphisms) that affected DNA cutting by one of the many just discovered "restriction enzymes" that cut DNA molecules within very specific base sequences.

Soon after the existence of DNA polymorphisms became known, proposals were made that they could provide the genetic markers needed to put together human genetic maps. In a 1980 paper, David Botstein, Ron Davis, Mark Scolnick, and Ray White argued that human maps could be obtained through studying the pattern through which polymorphisms were inherited in the members of large multigenerational families. Those polymorphisms that stay together were likely to be located close to each other on a given chromosome. During the next 5 years, two groups, one led by Helen Donis-Keller in Massachusetts, the other led by Ray White in Utah, rose to this challenge, both using DNA from family blood samples stored at CEPH (Centre d'Étude du Polymorphisme Humain), the mapping center established in Paris by Jean Dausset. By 1985, the mutant genes responsible for Huntington's disease and cystic fibrosis (CF) had been located on chromosomes 4 and 7, respectively.

By using a large number of additional polymorphic markers in the original chromosome 7 region implicated in CF, Francis Collins' group in Ann Arbor and L.C. Tsui's group in Toronto located the DNA segment containing the responsible gene. Its DNA sequence revealed that the CF gene coded for a large membrane protein involved in the transport of chloride ions. The first CF mutant they found contained three fewer bases than its normal equivalent and led to a protein product that was nonfunctional because of its lack of a phenylalanine residue.

The Human Genome Project: Responding to the Need for Efficient Disease Gene Mapping and Isolation

Although the genes responsible for cystic fibrosis and Huntington's disease were soon accurately mapped using only a small number of DNA polymorphic markers, the genes behind many other important genetic diseases quickly proved to be much harder to map to a specific chromosome, much less assign to a DNA chromosomal segment short enough to generate hopes for its eventual cloning. All too obviously, the genes behind the large set of still very badly understood diseases like Alzheimer's disease, late-onset diabetes, or breast cancer would be mapped much, much sooner if several thousands more newly mapped DNA polymorphisms somehow became available. Likewise, the task of locating the chromosomal DNA segment(s) in which the desired disease genes reside would be greatly shortened if all human DNA were publicly available as sets of overlapping cloned DNA segments (contigs). And the scanning of such DNA segments to look for mutationally altered base sequences would go much faster if the complete sequence of all the human DNA were already known. However, to generate these importantly new resources for human genetics, major new sources of money would be needed. So, by early 1986, serious discussions began as to how to start, soon, the complete sequencing of the 3×10^9 base pairs that collectively make up the human genome (the Human Genome Project or HGP).

Initially, there were more scientific opponents than proponents for what necessarily would be biology's first megaproject. It would require thousands of scientists and the consumption of some $3 billion-like sums. Those disliking its prospects feared that, inevitably, it would be run by governmental bureaucrats not up to the job and would employ scientists too dull for assignment to this intellectually challenging research. Out of many protracted meetings held late in 1986 and through 1987, the argument prevailed that the potential rewards for medicine as well as for biological research itself would more than compensate for the monies the Human Genome Project would consume during the 15 years then thought needed to complete it. Moreover, completion of each of the two stages—the collection of many more mapped DNA markers and the subsequent ordering of cloned DNA segments into long overlapping sets (contids)—would by themselves greatly speed up disease gene isolation.

Always equally important to point out, the 15 years projected to complete the Human Genome Project meant that its annual cost of $200 million at most would represent only 1–2% of the money spent yearly for fundamental biomedical research over the world. There was also the realization that some 100,000 human genes believed sited along their chromosomes would be much easier to find and functionally understand if genome sequences were first established for the much smaller, well-studied model organisms such as *Escherichia coli, Saccharomyces cerevisiae, Caenorhabditis elegans*, and *Drosophila melanogaster*. Thus, the biologists who worked with these organisms realized that their own research would be speeded up if the Human Genome Project went ahead.

The American public, as represented by their congressional members, proved initially to be much more enthusiastic about the objectives of the Human Genome Project than most supposedly knowledgeable biologists, with their parochial concerns for how federal monies for biology would be divided up. The first congressionally mandated monies for the Human Genome Project became available late in 1987, when many intelligent molecular geneticists still were sitting on the fence as to whether it made sense. In contrast, Congress, being told that big medical advances would virtually automatically flow out of genome knowledge, saw no reason not to move fast. In doing so, they temporarily set aside the question of what human life would be like when the bad genes behind so many of our major diseases were found. Correctly, to my mind, their overwhelming concern was the current horror of diseases like Alzheimer's, not seeing the need then to, perhaps prematurely, worry about the dilemmas arising when individuals are genetically shown at risk for specific diseases years before they show any symptoms.

Genome Ethics: Programs to Find Ways to Ameliorate Genetic Injustice

The moment I began in October 1988 my almost 4-year period of helping lead the Human Genome Project, I stated that 3% of the NIH-funded component should support research and discussion on the ethical, legal, and social implications (ELSI) of the new resulting genetic knowledge. A lower percentage might be seen as tokenism, while I then could not see wise use of a larger sum. Under my 3% proposal, some $6 million (3% of $200 million) would eventually be so available, a much larger sum than ever before provided by our government for the ethical implications of biological research.

In putting ethics so soon into the genome agenda, I was responding to my own personal fear that all too soon critics of the Genome Project would point out that I was a representative of the Cold Spring Harbor Laboratory that once housed the controversial Eugenics Record Office. My not forming a genome ethics program quickly might be falsely used as evidence that I was a closet eugenicist, having as my real long-term purpose the unambiguous identification of genes that lead to social and occupational stratification as well as to genes justifying racial discrimination. So I saw the need to be proactive in making ELSI's major purpose clear from its start—to devise better ways to combat the social injustice that has at its roots bad draws of the genetic dice. Its programs should not be turned into public forums for debating whether genetic inequalities exist. With imperfect gene copying always the evolutionary imperative, there necessarily will always be a constant generation of the new gene disease variants and consequential genetic injustice.

The issues soon considered for ELSI monies were far-ranging. For example, how can we ensure that the results of genetic diagnosis are not misused by prospective employers or insurers? How should we try to see that individuals know what they are committing themselves to when they allow their DNA to be used for genetic analyzing? What concrete steps should be taken to ensure the accuracy of genetic testing? And when a fetus is found to possess genes that will not allow it to develop into a functional human being, who, if anyone, should have the right to terminate the pregnancy?

From their beginnings, our ELSI programs had to reflect primarily the needs of individuals at risk of the often tragic consequences of genetic disabilities. Only long-term harm would result in the perception of genetics as an honest science if ELSI-type decisions were perceived to be dominated either by the scientists who provided the genetic knowledge or by the government bodies that funded such research. And because women are even in the distant future likely to disproportionately share the burden of caring for the genetically disabled, they should lead the discussions of how more genetic knowledge is to come into our lives.

Human Hesitations in Learning Their Own Genetic Fate

With the initial distribution of American genome monies and the building and equipping the resulting genome centers taking 2 years, the Human Genome Project in its megaphase did not effectively start until the fall of 1990. Decisions to go ahead by funding bodies in the United States helped lead to the subsequent inspired creation of Genethon outside Paris by the French genetic disease charity, Association Française contre les Myopathies (AFM), as well as the building of the now immense Sanger Centre, just south of Cambridge, England, by the British medically oriented charity, the Wellcome Trust. Now effectively 7 years into its projected 15-year life, the Human Genome Project has more than lived up to its role in speeding up genetic disease mapping and subsequent gene cloning. It quickly made successful the search for the gene behind the Fragile X syndrome that leads to severe mental retardation in boys preferentially affected by this sex-linked genetic affliction. The molecular defect found was an expansion of preexisting three-base repetitive sequences that most excitingly increase in length from one generation to the next. The long mysterious phenomenon of anticipation, in which the severity of a disease grows through subsequent generations, was thus given a molecular explanation. Then at long last, in 1994, the gene for Huntington's disease was found. Its cause was likewise soon found to be the expansion of a repetitive gene sequence.

While the mapping to a chromosome per se of any disease gene remains an important achievement, the cloning of the disease gene itself is a bigger milestone. Thus, the 1990 finding by Mary Claire King that much hereditary breast cancer is due to a gene on chromosome 17 set off a big gene-cloning race. With that gene in hand, there was a chance that its DNA sequence would reveal the normal function of the protein it codes for. In any case, it gives its possessors the opportunity to examine directly the DNA from individuals known to be at risk for a disease to see whether they had the unwanted gene. Thus, when in 1993 the chromosome 17 breast cancer gene (BRCA1) was isolated by Myriad, the Utah disease-gene-finding company, it could inform women so tested for BRCA1 whether or not they had the feared gene.

Initially, concerns were voiced that unbridled commercialization of this capability would all too easily give women knowledge they would not be psychologically prepared to handle. If so, the ethical way to prevent such emotional setbacks might be to regulate both how the tests were given and who should be allowed to be tested. I fear, however, that a major reason behind many such calls for regulation of genetic testing is the hidden agenda of wanting to effectively stop widespread genetic testing by making it so difficult to obtain. Now, however, calls for governmental regulation may fall on increasingly deaf ears. To Myriad's great disappointment, it appears that the great majority of women at 50% risk of being breast cancer gene carriers don't want to be tested. Rather than receive the wrong verdict, they seem to prefer living with uncertainty. Likewise, a very large majority of the individuals at risk for Huntington's disease are also psychologically predisposed against putting themselves at risk of possibly knowing of their genetic damnation.

Although we are certain to learn in the future of many individuals regretting that they subjected themselves to genetic tests and wishing they had been more forewarned of the potential perils of such knowledge, I do not see how the state can effectively enter into such decisions. Committees of well-intentioned outsiders will never have the intimate knowledge to assess a given individual's psychological need, or not, for a particular piece of scientific or medical knowledge. In the last analysis, we should accept the fact that if scientific knowledge exists, individual persons or families should have the right to decide whether it will lead to their betterment.

Inarguable Existence of Genes Predisposing Humans to Behavioral Disorders

The extraordinarily negative connotations that the term eugenics now conveys are indelibly identified with its past practitioners' unjustified statements that behavioral differences, whether between individuals, families, or the so-called races, largely had their origins in gene differences. Given the primitive power of human genetics, there was no way for such broad-ranging assertions to have been legitimatized by the then-current methods of science. Even the eugenically minded psychiatrists' claims that defective genes were invariably at the root of their mental patients' symptoms were no more than hunches. Yet, it was by their imputed genetic imperfection that the mentally ill were first sterilized and then, being of no value to the wartime Third Reich, released from their lives by subsequent "mercy killings."

But past eugenic horrors in no way justify the "not in our genes" politically correct outlook of many left-wing academics. They still spread the unwarranted message that only our bodies, not our minds, have genetic origins. Essentially protecting the ideology that all our troubles have capitalistic exploitative origins, they are particularly uncomfortable with the thought that genes have any influence on intellectual abilities or that unsocial criminal behavior might owe its origins to other than class or racially motivated oppression. However, whether these scientists on the left actually believe, say, that the incidence of schizophrenia would seriously lessen if class struggles ended is not worth finding out.

Instead, we should employ, as fast as we can, the powerful new techniques of human genetics to find soon the actual schizophrenia predisposing genes. The much higher concordance of schizophrenia in identical versus nonidentical twins unambiguously tells us that they are there to find. Such twin analysis, however, reveals that genetics cannot be the whole picture. Because the concordance rates for schizophrenia, as well as for manic–depressive disease, are more like 60%, not 100%, environmental predisposing factors must exist and, conceivably, viral infections that affect the brain are sometimes involved.

Unfortunately, still today, the newer statistical tricks for analyzing polymorphic inheritance patterns have not yet led to the unambiguous mapping of even one major schizophrenic gene to a defined chromosomal site. The only convincing data involve only the 1% of schizophrenics whose psychoses seemingly are caused by the small chromosome 22 deletions responsible also for the so-called St. George facial syndrome. Manic–depressive disease also has been more than hard to understand genetically. Only last year did solid evidence emerge for a major predisposing gene on the long arm of chromosome 18. This evidence looks convincing enough for real hopes that the actual gene involved will be isolated over the next several years.

Given that over half the human genes are thought to be involved in human brain development and functioning, we must expect that many other behavioral differences between individuals will also have genetic origins. Recently, there have been claims that both "reckless personalities" and "unipolar depressions" associate with specific polymorphic forms of genes coding for the membrane receptors involved in the transmission of signals between nerve cells. Neither claim now appears to be reproducible, but we should not be surprised to find some subsequent associations to hold water. Now anathematic to left-wing ideologues is the highly convincing report of a Dutch family, many of whose male members display particularly violent behavior. Most excitingly, all of the affected males possess a mutant gene coding for an inactive form of the enzyme monoamine oxidase. Conceivably having too little of this enzyme, which breaks down neurotransmitters, leads to the persistence of destructive thoughts and the consequential aggressive patterns. Subsequent attempts to detect in other violent individuals this same mutant gene have so far failed. We must expect someday, however, to find that other mutant genes that lead to altered brain chemistry also lead to asocial activities. Their existence, however, in no way should be taken to mean that gene variants are the major cause of violence. Nonetheless, continued denials by the scientific left that genes have a role in how people interact with each other will inevitably further diminish their already tainted credibility.

Keeping Governments Out of Genetic Decisions

No rational person should have doubts whether genetic knowledge properly used has the capacity to improve the human condition. Through discovering those genes whose bad variants make us unhealthy or in some other way unable to function effectively, we can fight back in several different ways. For example, knowing what is wrong at the molecular level should let us sometimes develop drugs that will effectively neutralize the harm generated by certain bad genes. Other genetic disabilities should effectively be neutralized by so-called gene therapy procedures restoring normal cell functioning by adding good copies of the missing normal genes. Although gene therapy enthusiasts have promised too much for the near future, it is difficult to imagine that they will not with time cure some genetic conditions.

For the time being, however, we should place most of our hopes for genetics on the use of antenatal diagnostic procedures, which increasingly will let us know whether a fetus is carrying a mutant gene that will seriously proscribe its eventual development into a functional human being. By terminating such pregnancies, the threat of horrific disease genes continuing to blight many families' prospects for future success can be erased. But even among individuals who firmly place themselves on the pro-choice side and do not want to limit women's rights for abortion, opinions frequently are voiced that decisions obviously good for individual persons or families may not be appropriate for the societies in which we live. For example, by not wanting to have a physically or mentally handicapped child or one who would have to fight all its life against possible death from cystic fibrosis, are we not reinforcing the second-rate status of such handicapped individuals? And what would be the consequences of isolating genes that give rise to the various forms of dyslexia, opening up the possibility that women will take antenatal tests to see if their prospective child is likely to have a bad reading disorder? Is it not conceivable that such tests would lead to our devoting less resources to the currently reading-handicapped children whom now we accept as an inevitable feature of human life?

That such conundrums may never be truly answerable, however, should not concern us too much. The truly relevant question for most families is whether an obvious good to them will come from having a child with a major handicap. Is it more likely for such children to fall behind in society or will they through such affliction develop the strengths of character and fortitude that lead, like Jeffrey Tate, the noted British conductor, to the head of their packs? Here I'm afraid that the word handicap cannot escape its true definition—being placed at a disadvantage. From this perspective, seeing the bright side of being handicapped is like praising the virtues of extreme poverty. To be sure, there are many individuals who rise out of its inherently degrading states. But we perhaps most realistically should see it as the major origin of asocial behavior that has among its many bad consequences the breeding of criminal violence.

Thus, only harm, I fear, will come from any form of society-based restriction on individual genetic decisions. Decisions from committees of well-intentioned individuals will all too often emerge as vehicles for seeming to do good as opposed to doing good. Moreover, we should necessarily worry that once we let governments tell their citizens what they cannot do genetically, we must fear they also have power to tell us what we must do. But for us as individuals to feel comfortable making decisions that affect the genetic makeups of our children, we correspondingly have to become genetically literate. In the future, we must necessarily question any government that does not see this as its responsibility. Thus, will it not act because it wants to keep such powers for itself?

The Misuse of Genetics by Hitler Should Not Deny Its Use Today

Those of us who venture forth into the public arena to explain what genetics can or cannot do for society seemingly inevitably come up against individuals who feel that we are somehow the modern equivalents of Hitler. Here we must not fall into the absurd trap of being against everything Hitler was for. It was in no way evil for Hitler to regard mental disease as a scourge on society. Almost everyone then, as still true today, was made uncomfortable by psychotic individuals. It is how Hitler treated German mental patients that still outrages civilized societies and lets us call him immoral. Genetics per se can never be evil. It is only when we use or misuse it that morality comes in. That we want to find ways to lessen the impact of mental illness is inherently good. The killing by the Nazis of the German mental patients for reasons of supposed genetic inferiority, however, was barbarianism at its worst.

Because of Hitler's use of the term *Master Race*, we should not feel the need to say that we never want to use genetics to make humans more capable than they are today. The idea that genetics could or should be used to give humans power that they do not now possess, however, strongly upsets many individuals first exposed to the notion. I suspect that such fears in some ways are similar to concerns now expressed about the genetically handicapped of today. If more intelligent human beings might someday be created, would we not think less well about ourselves as we exist today? Yet anyone who proclaims that we are now perfect as humans has to be a silly crank. If we could honestly promise young couples that we knew how to give them offspring with superior character, why should we assume they would decline? Those at the top of today's societies might not see the need. But if your life is going nowhere, shouldn't you seize the chance of jump-starting your children's future?

Common sense tells us that if scientists find ways to greatly improve human capabilities, there will be no stopping the public from happily seizing them.

Five Days in Berlin

1997

In late April, when flying off to Berlin to give the keynote talk before a
German organized Congress of Molecular Medicine, I was apprehensive that my forthcoming message on *Genes and Politics* was not what
my hosts had bargained for in wanting a genetic celebrity on hand to mark
Germany's long-delayed return to a prominent role in human genetics.
Great Britain and France as well as the United States had now for more
than a decade been leading players in the Human Genome Project, the
worldwide effort to determine the sequence of the 3×10^9 bits of DNA
information encompassed within the 24 different human chromosomes
(the human genome). Several years later Japan, always beset by the need
for a time-consuming, broad consensus before accepting a new challenge,
also opted to become a major participant. So among the major technologically advanced nations, it has been only Germany which up until now has
played no significant role in this profoundly exciting effort to vastly speed
up our knowledge of the human genetic information.

In one sense, this German reluctance to join was not to be expected
because it was obvious that medical research would be immeasurably
speeded up by the finding of those genes which, when miscopied during
DNA replication, lead to human diseases like cancer, diabetes, or
Alzheimer's. In having no genome program, German scientists until now
have been virtually excluded from current disease-gene-finding successes
and their valuable patentable consequences. Secondly, those pursuing the
Human Genome Project invariably will be generating new DNA methodologies pivotal for future advances in biotechnology, that is, the use of cells
for generating new commercial products (drugs, diagnostics, foods, etc.).

Major nations that stay away from the human genome are bound to imperil their long-term commercial futures.

Germany's absence from the genome table thus never reflected financial considerations. Instead, politics were involved. Germany's past involvement with eugenics ("genetics used for human improvement") was a moral disaster in which all too many of their leading human genetics practitioners eugenetically preached racial Nordic superiority and willingly participated in the 1933–1945 Nazi era elimination by scientific selection of Germany's mentally ill, Jews, and Gypsies. After the war ended and these atrocities became known, their most direct perpetuators were tried under the Nuremberg Laws and, among those that did not commit suicide, a number were executed. However, those academics whose hands were not so directly bloodied and who claimed to never have been more than scientific advisers slowly crept back into leading academic positions in genetics, psychiatry, and anthropology. The German nation thus never directly faced up to the moral depravity perpetrated under the name of genetics. Far better would have been an effective moratorium on the teaching of these subjects for a decade or two following the war. Then a new generation of teachers trained abroad could go back to Germany. Instead the rot of Nazi genetics tainted the German university system until the late 1960s.

So neither genetics nor geneticists had the smell of integrity in Germany when the recombinant DNA era began in the 1970s and opened up the possibility of gene cloning and genetic manipulation. That this was not a new era we should blindly rush into was immediately realized by the DNA community that quickly instituted an effective self-imposed moratorium until serious consideration was given to whether the novel genetically modified microorganisms (viruses, bacteria, and yeast) we would create in our laboratories might pose a realistic threat to either human health or the world's ecology. A large international meeting was convened near Monterey, California, to discuss these issues that later became the province of specialist-dominated governmental committees formed in those countries whose scientists wanted to use recombinant DNA procedures for either understanding phenomena like cancer or the immunological response or to create genetically engineered cellular factories for the making of important new pharmaceutical drugs or diagnostics. Virtually all such committees, independent of national origin and usually containing members drawn from the general public, opted by the end of 1979 to effectively end the moratorium for most research and biotechnology uses.

Though at first there were discussions that the public would be more reassured by the creation of laws as opposed to government-agency-promoted regulations, the inherent complexity of writing laws to handle the extraordinarily broad ways in which recombinant DNA might be employed always made more easily changeable government regulation the preferred way to handle the fast-moving pace of innovation in recombinant DNA use. Partly because of ever-decreasing regulations, recombinant DNA-based research has thrived in the Western Hemisphere, in Japan, and in most parts of Europe. Here the major exception not surprisingly has been Germany, all too long accustomed to viewing geneticists as individuals more responsive to the needs of themselves or the state as opposed to the public good.

For almost two decades the totally unjustified German public perception that recombinant DNA manipulation per se is inherently evil has led to much unnecessary, if not ludicrous, regulation and legislation that has greatly set back their country's biological and medical research as well as long stifling the use of DNA technologies by German industry, be it large or small. Instead of analyzing individual RNA recombinant products for their potential risks to humans or to the broader issue of the world ecology, German DNA opponents have misleadingly raised the moral threat of Nazi-type geneticists working to overturn the established natural order of life. In letting hysteria triumph over the use of reason, recombinant DNA products like insulin cannot be industrially produced in most parts of Germany. As a result, there has been an exodus of much German pharmaceutical research to other countries. Likewise the unwarranted belief that any plant modified by a recombinant DNA procedure is inherently dangerous and never should be grown outside the protective custody of a greenhouse has resulted in the vandalized destruction of each of 14 attempts to grow such plants in open German fields last summer. And not only the labs but also the homes of prominent German scientists interested in human genetics have been threatened by terrorist bombs.

More concerned that they be perceived by their citizens as seeming to do good than in actually doing good, the German government, never coming to grips with why the anti-genetics feeling remained so high among its people, long sat on its hands unrealistically hoping that the anti-DNA feeling would die down. If the German economy had remained in high gear, I suspect that both biotechnology and human genetics research would even today remain stifled. But with the massive economic readjustment necessary to bring the former East German economy back to life, a now worried German government two years ago finally had the sense to initiate a modest Human Genome Project as well as to finally jumpstart German biotechnology through federal and state subsidies.

Knowing of this most welcome turnabout, a year ago I accepted an invitation to come a year hence to Berlin for a meeting. No longer would my hosts have to be inherently gloomy at effectively being kept out of the ever-growing excitement of today's human genetics and biotechnology. My main reason for going, however, was that the occasion was sponsored by the Max Delbrück Center for Molecular Medicine, a large research institution created after German unification in Berlin-Buch on the grounds of a former East German Institute for Biological Research. Even earlier it was the site of important 1930s research on the gene by the highly talented Russian geneticist Nicolai Timofeeff Ressovsky, collaborating with the young German theoretical physicist Max Delbrück, whose interests were switching from physics to biology. Strongly influenced several years earlier in Copenhagen by Niels Bohr, Max worked in Berlin-Dahlem at The Kaiser Wilhelm Institut fur Chemie with the famed Lisa Meitner. Later as a refugee in Stockholm, she with her nephew Otto Frisch correctly interpreted the uranium fission process discovered in her former institute by Otto Hahn and Fritz Strassmann.

In 1935, Delbrück and Timofeeff wrote a seminal paper that attempted to measure the size of the gene through the rate at which it inactivated by increasing doses of X rays. Their approach, given the name "Target Theory," became widely known later through Erwin Schrödinger's 1944 book, *What Is Life?* This little but vastly influential book appeared after a series of lectures he delivered in Dublin, to which he had gone as a refugee from Nazi Austria. Its discussion of the Delbrück model of the gene first caught my attention in the late 1940s when I was a biology student at the University of Chicago. As a result, I stopped thinking of birds as my future career and instead saw the nature of the gene as biology's most important objective.

By then Max Delbrück, whose high Protestant academic background proved immiscible with a Nazi indoctrination weekend, was in the United States, moving in 1937 to the genetics world at the California Institute of

Technology. Soon after his arrival in Pasadena, Max hit upon bacterial viruses (phages) as perfect experimental systems for the gene studies and soon set into motion incisive research which over the next decade led to his 1969 Nobel Prize. My getting to know him during the summer of 1948 at the Cold Spring Harbor Laboratory was a revelation, with his tall youthful esprit soon becoming the model for what I wanted out of my own life. How one gene becomes two identical genes was the problem to solve. When Francis Crick and I found the answer through our March 1953 discovery of the double helical structure of DNA, Max instantly rose to the occasion, proclaiming to all that the double helix would be the starting point for all future biological research. From him I had learned that Dahlem, before Hitler, had been one of the world's great scientific centers, and I now looked forward to seeing what luster, if any, from the past still enveloped the Dahlem scene.

During a day's stopover in Oxford before going on to Berlin, I began to worry more whether I would soon be unnecessarily complicating my life. Instead of praising my hosts for finally starting a Human Genome Program, I would be telling them that the world wants them to finally shape up about their Nazi genetic past. I could do so in part because I was not making them responsible for all past genetics misdeeds. A prominent eugenics movement also existed in the United States before, between 1910 and 1940. It was centered in fact at the Cold Spring Harbor Laboratory, where I had first gotten to know Delbrück and of which since 1968 I have been Director and now President. Like the German effort, the American Eugenics agenda was primarily political as opposed to scientific and irresponsibly preached prejudice as scientific fact.

To restrict the level of Italian immigration into the United States, Harry Laughlin, the director of the Laboratory's Eugenics Record Office, gave testimony in 1923 before the American Congress that the people from Southern Europe were genetically prone to criminality. Furthermore, the equally unjustified belief of many then practicing human geneticists that race mixing leads to degenerate offspring helped pass laws in many American states that forbade interracial marriages. So I could start my talk with American misdeeds before going on to the clearly much viler German situation where, for example, the "incurable" psychiatric patients with supposed inherited defects were called "socially inferior, empty shells or ballast existences." So identified between January 1940 and September 1941, some

70,000 already sterilized psychiatric patients were murdered in gas chambers to free up their hospital beds for the war wounded.

Until the war started, not withstanding the anti-semitic Nuremberg decrees of 1935, the American and the German Eugenics movements remained enthusiastic partners, with Harry Laughlin receiving in New York from the German Consul an honorary degree from Heidelberg University for his promotion of racial hygiene. Even as late as 1938, at a *Eugenics Rally* in Pasadena, California, Laughlin praised the great benefit from the July 14, 1933 Nazi Law for the prevention of progeny with hereditary defects. It allowed for compulsory sterilization for 16 different supposed genetic defects including congenital mental defects, schizophrenia, manic-depressive disease, hereditary blindness and deafness, hereditary epilepsy, and severe alcoholism. Over the next six years, almost 400,000 persons so diagnosed were sterilized.

Though I never talk from manuscripts, in this case not wanting to be misquoted, I saw the need to have one prepared to hand out after my talk. Worried that I might have some facts wrong, just before getting on the plane I arranged for a copy to be faxed to my late 1960s friend at Harvard, Benno Müller-Hill, now a Professor of Genetics at Cologne and whose 1988 English translation of his 1984 book (*Todliche Wissenshaft*) *Murderous Science* had first alerted me to what had not happened to the Nazi co-opted geneticists after the war. To my relief, soon after getting to Berlin, Benno phoned me at the InterContinental Hotel that my facts about the past were right on the mark.

With a free day ahead of us, my wife Liz and I spent the morning in former East Berlin at the Pergamon Museum, famed for its almost intact Grecian temples and the extraordinary Assyrian brick gates from Babylon that several generations of German archaeologists sent back to Berlin beginning around 1870. In the afternoon, once back in the former West Berlin, we walked around the Prussian Kings' Berlin palace at Charlottenburg, noting in particular the relatively petite, two-floored "summer home" designed for Karl Frederick IV by Karl Frederick Schinkel, the great German classical architect of the early nineteenth century whom our late architect friend, Charles Moore, much admired. Schinkel's English equivalent, Sir John Soane, was also a Moore favorite; and the Soane-like features of *Ballybung*, our new home at Cold Spring Harbor, would not have been out of place in Schinkel's "summer house."

For the early evening our host Detlev Ganten, the Director of the Max Delbrück Center, arranged for a pre-dinner visit to the former atelier of Jeanne Mammen (1890–1976), the Berlin-born painter whom Max formed a close friendship with just before he left for the United States. Until the Nazis took over, she was a successful magazine illustrator, drawing and painting men and women of the 1920s Berlin with a George Grosz-type sting. Such drawings became impossible to do with Hitler in power, and during the war her painting had Picasso-like features. After the war, Max brought back several of these 1940s pieces to his Caltech home; here they dominated the big room where Max played the piano and entertained the always continuous stream of young guests that he and his wife Manny were partial to.

Mammen's atelier was on the top floor of a large early twentieth-century building on the Kurfurstendamm and now maintained as a mini-museum by several friends who helped look after her in her last years. With Max 23 years older than I, he had once had a father-like role to me. So with Ganten following me out of breath, I bounded up the staircase imagining I was still the 20-year-old youth in awe of everything that was Delbruck. It was my hope that I could somehow buy a 1920s drawing from the Jeanne Mammen Gessellschaft, but soon found they had largely found their way to museums like the Busch Reisinger at Harvard. Instead I happily settled upon a bronze statue (from the war years) of a head that I took to be a haunted soldier of the Third Reich. With other guests wondering whether it might have been modeled upon memories of Max, I silently noted its resemblance to an oil of a wartime soldier that hung in the next room.

The occasion had started by a short speech in German, with Ganten sitting between us on the couch whispering in English into our ears. Before coming to Berlin, I only thought of how unique Max had been to me, but never vice versa. So I felt awkwardly happy when Max's feelings about me were highlighted, as well as his just post-double-helix notion that I was the Einstein of Biology. Max used to emphasize that both Albert Einstein and Werner Heisenberg were only 25 when they did their best science, leading me later to take satisfaction in the fact that I was one month short of being 25 when Francis Crick and I found the double helix. Such feelings, however, always were accompanied by the realization that our finding of the double helix did not represent a particularly difficult intellectual effort. But now before new acquaintances not in the know I only smiled and accepted the champagne that was brought out to mark our still warm memories of Max.

Initially I was attracted to giving the keynote speech by knowledge that it was to be preceded by a talk from the German Science Minister. So my arrival in Berlin was clouded by learning that a more junior representative, Elke Wulfing, was to give the Bonn government presentation. Her speech, nonetheless, proved most informative for what she said as well as what she didn't say. Emphatically the driving force behind Germany's belated entry into the genome game was economic—to directly help its pharmaceutical companies as well as to push forward Germany's fledgling biotechnology efforts. Nonetheless, she gave no encouragement that the 24 million dollars now allocated for genome efforts would soon grow. And when she said that research that didn't lead to patents had no place in her research portfolio, virtually everyone turned their stunned heads to catch the reaction of their compatriots. I fear that her desire for immediate commercial research payoffs represents the thinking of most equivalent governmental ministers elsewhere.

All in all, I got the unmistakable impression that Germany was now only getting into the genome game to help itself as opposed to the world as a whole. Immediately explainable was why Germany was not going along with the unselfish data release policies agreed on more than a year ago by United States and United Kingdom scientists whereby they put all their genome data on the Worldwide Web within a day of generation. In contrast, the German government now wishes to give its industry a 90-day first look at German-generated genome data before letting the rest of the world see them. In so going alone and not being part of the now very efficient USA–UK collaborative programs, Germany's genome effort may not turn out to be the locomotive its government wants to pull its biotechnology industry into world prominence. My feeling of angry despair was compounded when her talk ended without any significant mentioning of the ethical uses of genetic information.

With my adrenalin levels then clearly raised, I improvised the starting words of my own talk to say that Ms. Wulfing should report back to Bonn that the outside world saw little to cheer about in the way Germany was entering the genome world. This was not the occasion for her to portray genes as potential servants of the commercial interests of the German state. Instead she should be emphasizing, in particular before a Congress of Molecular Medicine, the people of the world's desire to find ways to combat the pernicious effects of gene-related diseases. My necessary scorn directed toward Bonn brought forth enthusiastic applause by the audience, almost all connected to medically related research in Germany.

Later it was much harder to judge their reactions to my many slides illustrating what eugenics really stood for prior to World War II. Certainly the squalid American story to which most of my talk was devoted was new to their eyes. And I suspect few of them had been consciously exposed to the grisly details of how the German eugenicists gave whole-hearted support to the Nazi doctrine of race purity and their program to exclude anything Jewish from future German life. At the end of my hour I emphasized how bad it had been for the postwar German government to bring back into German academic life Professors Fritz Lenz and Otmar von Verschuer. Both had continued working in behalf of the Nazis even after they knew the genocidal final solution to the Jewish problem had commenced. Here I had to say that the 1939 Nobel Prize winner, Adolf Butenandt, should be most remembered for his participation in the 1949 whitewash of Professor von Verschuer, then well known for wartime research using materials sent to him from Auschwitz by his former student, Josef Mengele. Von Verschuer, mildly reprimanded for "a few isolated events of the past," afterwards became the Professor of Genetics at Munster while Butenandt went on to lead the Max-Planck-Gesellschaft.

I had to pose the question whether anyone really should be surprised that so many Germans still mistrust their geneticists if not their entire scientific community. Had not the time come for the German genetics community to finally admit that they are deeply ashamed of their forebears' past? In my mind, only by doing so will it stand the chance of finally escaping from the ghost of Hitler that still haunts them as well as geneticists all over the world. We don't want Hitler's name brought up over and over as we try to somehow mitigate the genetic inequalities that diminished so many innocent peoples' lives.

Though there was much applause from the audience of over 1500 scientists, I had to wonder whether my true message got across. To most who came up to me, I gave copies of my written manuscript, particularly to reporters who wanted one-minute-type quotes that by then I was too exhausted to give. By attacking the integrity of Butenandt, I knew I would go deep into the German psyche, but it was his actions, not Hitler's, that are Germany's problem now. To my relief, Detlev Ganten turned warmly to me, telling me he was glad that I brought up the issue that Germany has never really purged itself of the sins of its Nazi-era geneticists. Then soon I was being congratulated by Peter Fischer, the biographer of Max Delbrück and now Professor of the History of Science at Konstanz University.

Later at dinner at the Funkturm Restaurant, on top of the radio tower next to the Congress Hall on Berlin's outskirts, I sat next to Elke Wulfing, telling her about how our American genome effort was committing five percent of its funds to discussion of the ethical, legal, and social consequences of genome research. Telling me her field of past expertise was textiles, she emphasized that her ministry had delayed starting its genome effort until they sorted out the ethical issue. In our congress packet was a March 1997 pamphlet from her Ministry of Science and Technology entitled *Biotechnology, Genetic Engineering and Economic Innovation*. In it was the statement that there was need for interdisciplinary research into ethical, legal, and social questions. Nowhere within it did I find the word eugenics mentioned, nor was there any discussion of why Germany faces greater difficulty than any other nation in pursuing the study of human DNA. As the dinner ended, I got the unmistakable impression that Ms. Wulfing thought genome ethical issues were now at last behind the backs of her ministry.

The next afternoon, Liz and I were given a car and driver to take us out to Dahlem, where Max had grown up, to see the prewar buildings of the Kaiser Wilhelm Institutes. To help us, Detlev Ganten arranged for us to be guided by a professor interested in prewar medicine. Upon picking him up at a nearby suburb, I explained that I was particularly interested in seeing the building that once housed the Kaiser Wilhelm Institute for Anthropology, Human Genetics, and Eugenics. It had been created especially for Professor Eugen Fischer, the leader of the German eugenic movement who for more than 20 years was in close correspondence with Charles B. Davenport, the founder of Cold Spring Harbor's 1910 Eugenics Record Office. Then Fischer's worldwide reputation, gained in part through his studies on race mixing in German Southwest Africa (Nambia), gained him funds from the Rockefeller Foundation to cover his building's 1926 construction. Later, as the 1933–1934 rector of Berlin University after Hitler came to power, Fischer signed the dismissal notes of its Jewish professors.

Officially retiring in 1942, Fischer was replaced as Institute Director by Otmar von Verschuer, who five years before had moved from Berlin to Frankfurt as Professor of Genetics. In 1937, von Verschuer, in a letter to Fischer, mentioned his report to Alfred Rosenberg (later the Reich Minister of the Occupied Eastern Territories) containing his proposals for the registration of Jews and half-Jews. Later that year von Verschuer protested to the Reich Minister of Justice Gurtner that his expert opinion (written with the help of Josef Mengele) incriminating the defendant in a race dishonor trial (marriage of Jews with Aryans was then forbidden) had not been accepted and that as a result the defendant had been set free.

After looking up at the turreted, four-story building where Hahn and Strassman in 1939 had split the uranium atom, we walked toward the KWI for Anthropology building with our guide explaining that its last professor had been Professor von Verschuer, who after the war had moved on to the University of Munster from which he had traveled to an international meeting in Venezuela to talk about his twin research. Sensing that he had not been at my lecture, I quietly let him discover that von Verschuer's brand of genetics was not my type. So he knew what I might be thinking when we came upon a bronze plaque next to the front door of the two-story, some 400-foot-long KWI building. Its 21 lines described the past history of the building and how its science had been perverted for political purposes. After the war, this building became part of the Frei University of West Berlin. With its lower floor bracketed by New York subway-type graffiti that now cover too many Berlin buildings, social scientists now make it their home. In leaving it, I thought how much better for it to have been bulldozed to the ground immediately after the war.

In later passing first the KWI building once housing the preeminent biochemist Otto Warburg and later the building built for Einstein before he saw good reason not to return to Berlin, I brought up the awfulness of racism, thinking of how grand Dahlem had been when it welcomed Jews among its institutes. Our guide, however, not understanding my words, quickly voiced agreement, saying that with today's bad unemployment, racism was again a real feature of Berlin existence.

With our guide now gone, Liz and I drove further out to Lake Wannsee and across the bridge where spies could be exchanged between East and West Berlin, to the once grand Potsdam in former East Germany which still looked as if World War II had just ended. There we first walked about Cecilianhof, the last Kaiser's eldest son's Lutyens-style great lakeside 1916 mansion where Stalin, Churchill, and Roosevelt met just after Europe was liberated. Then on to Sans Souci, the Prussian House of Hohenzollerns' several-mile-square grand estate of palaces and gardens. In admiring its yellow-gilded Chinoise teahouse, its sprawling Italianate bath house, and its Charlottenhof elegant blue-windowed, stone Schinkel mini-palace of the 1830s, I appreciated better a key fact told to us soon after our arrival in Berlin. Until this visit I had assumed that Berlin's becoming the capital of the 1871 unified Germany was a successful reflection of Prussian militarism. Instead we were told that Berlin's preeminence had come about in large part because the eighteenth-century Prussian kings welcomed immigrants and new ideas, eagerly accepting the Huguenots forced out of eighteenth-century France. Likewise Berlin, more than any other German city, gave its Jews full rights allowing them to become by 1900 highly prominent members of its academic and commercial worlds. As a result, many Jews fought valiantly on the Kaiser's side in the First World War. It was Prussian enlightenment, not military might, that allowed Berlin to become perhaps the world's important cultural center for the first third of this century.

A free last day in Berlin gave us the opportunity for the 45-minute drive north to Berlin-Buch to see how the newly created Max Delbrück Center for Molecular Medicine was faring. In driving us out, Detlev Ganten wanted us to walk about an adjacent set of some 20 hospital buildings that soon would be totally vacated. One possibility for their subsequent use would be as a conference center for continued medical education. The whole complex was almost a small town with all the buildings from gatehouse to the grand doctors' homes to the several hundred room hospital buildings all constructed from handsome red brick and topped by roundish Flemish gables. For a moment I felt that I might be in the midst of a large group of early twentieth-century American college dormitories. How the East Germans used these buildings did not seem to interest our hosts, with my getting the impression that tuberculosis patients once might have roamed its gardened grounds.

Then moving past a gated lodge into the Max Delbrück Center (MDC), where Detlev Ganten's office was on the second floor of the massive four-story, long research building. Its total reconstruction only par-

tially compensated for the shoddy East German construction of only 15 years ago. Also being converted for use by Max Delbrück scientists was the large brick-covered "1920s modern" building that once housed Oscar Vogt's prestigious KWI for Brain Research. In the mid 1920s, Vogt had been asked by Stalin to come to Moscow to examine Lenin's brain. Though Lenin's thinking parts proved anatomically similar to all others, the visit let Vogt meet the talented young geneticist, Timofeeff Ressovsky, whom he persuaded to join his Berlin-Buch Institute. Living in the large gatehouse that Ganten now hopes soon will be transformed into the institute's club house, Timofeeff, at the war's end, remained in Buch after virtually all its other scientists had fled to western parts of Germany. As such, he was soon captured by the Russian army and transported back to Russia. There he was imprisoned for many years until moved to a military research lab in the Ural Mountains, where he was an imprisoned scientist until his death in 1981.

By 1937 Oscar Vogt no longer could work under the Nazis and moved to a private lab in the Black Forest. At that time, Professor Julius Hallervorden from the Brandenburg Psychiatric Institute arrived in Buch. There he took particular interest in the brains of the feeble-minded that he obtained after they were "euthanized" by the Nazis. Much additional material came to him from the mental hospital at Brandenburg Görden that was closely connected to an extermination center in nearby Brandenburg. Hallervorden was actively involved with this center, not only providing it with a technician from his institute but also letting one of its physicians come to his lab to do research and learn advanced anatomical techniques. Yet after American interrogation he somehow avoided postwar imprisonment, becoming head of the Max Planck Institute of Brain Research in Giessen in 1948.

When our tour of the center was over and I had signed its new large guest book, we went for dinner at Detlev's new home built on land just over the boundary from the MDC in the state of Brandenburg. With his house cheerfully new and having a garden to step out into from the living room, it was a welcome relief to his physician wife, who lived for two years in the once-grand Oscar Vogt house after they arrived from Heidelberg. Admitting that her first reaction to life in the former East Germany was of deep despair, the large new complex of modern flats across the field into Brandenburg provided visible proof that East Germany was mending, albeit slowly. Dinner was dominated by conversation with two of the rare East German scientists kept on when the MDC was formed. They still read the newspaper of East as opposed to West Berlin, and they liked the way child care had been handled in East Berlin more than the system now taking over.

At dessert time, I raised the question of whether any state funds could be used to give the MDC a more distinguished visual appearance but learned that the bureaucrats do not allow such frivolities. A few years ago, German industry might have risen to the occasion. Now with their profits low, they are miserly even with monies for supporting research that may have industrial payoffs. Asking whether private philanthropy will soon take off in a big way, I was told that virtually all Germans, rich to poor, assume that the state takes care of all important tasks. With Prussian kings no longer to be counted on and Berlin's Jews reduced to at most a few thousand, the federal and state governments remain the only bodies people can count on.

When we got back to our hotel, Liz immediately opened the English-language version of the thick book, which we had bought at the massive Potsdamer Platz construction site, to read about the Berlin region's architectural riches. In the chapter dealing with Berlin-Buch, she found its write-up of the early twentieth-century Flemish gabled hospital buildings. As we suspected but did not earlier say, the Kaiser's 1906 government took great care of, and chose distinguished architects for, their lunatic asylums.

Ten days later, Germany ever haunts me.

Good Gene, Bad Gene: What Is the Right Way to Fight the Tragedy of Genetic Disease?

1997

Genes too often get a bad press. This is not surprising because there are "bad" genes as well as "good" ones, and bad news grips readers more than good news. Bad genes are actually mutated good genes that, because of altered DNA messages, do not function normally. One particularly bad gene leads to Huntington's disease, which progressively destroys key nerve cells. Most of an individual's genes, however, are inherently good. Collectively, they are the instruction book for our bodies: Without the right instructions from our genes, we could not develop into functioning adults. And fortunately, many bad genes—like that for cystic fibrosis—have no immediate consequence because they are expressed only when copies are inherited from both the father and mother. Carriers possessing only one copy of this gene are much more common (around one in 25) than individuals with the disease (around one in 2300).

Until recently, there was no way to isolate and characterize bad genes. They were known only by their consequences: disease. Today, however, thanks to the development of powerful new ways for studying DNA, there is a flood of information about the faulty genes implicated in virtually every major human disease, including diabetes, cancer, and asthma. Every week or so a new disease gene is discovered.

But with almost routine ways now available to test DNA samples for the presence of specific mutant genes, there is increased anxiety that an individual's genetic heritage may be vulnerable to unwanted prying. The

223

DNA from a single human hair, for example, may be sufficient to alert a prospective employer or health insurer to a person's genetic predisposition to disease. Broad privacy laws must therefore be enacted to forbid genetic tests without the informed consent of the individual involved. But even with such laws, dilemmas will arise when individuals do not realize the significance of the proposed genetic screening. These tests warn of impending disease, but do not cure. And how many people would want to have certain knowledge that they will contract a disease for which there is no cure?

Banishing genetic disability must therefore be our primary concern. We would not worry about testing for a predisposing gene for Alzheimer's disease if we already had the cure. In this case, knowing that an individual is seriously predisposed might allow drug therapy to begin before brain functioning is irreversibly diminished. The recent discovery of several genes whose malfunctioning leads to Alzheimer's disease provides the pharmaceutical industry with important molecular targets for drug development. Only through the discovery of these kinds of genes can biomedical research stop this most pernicious cause of human senility.

We must never, however, live under the misconception that we will ever effectively control the majority of genetic diseases. Many are likely to prove intractable to drug therapies or gene therapies, in which good genes are introduced into cells to compensate for bad ones. It will be particularly difficult to compensate for genes that malfunction during fetal development. If key genes controlling the networking of brain cells don't come into action in the womb, no drug or gene therapy procedure will be able to correctly rewire the brain later.

There is a great difference of opinion as to whether steps should be taken to prevent the birth of genetically impaired children. Many are opposed for religious reasons to trying to control the genetic destinies of children. Others, recalling Germany's eugenic practices, have an equally strong abhorrence of genetic-based reproductive decisions. These people fear a more powerful eugenic practice, whereby the crude racial and class prejudices of early eugenicists are replaced by scientific demonstrations of genetic inequality.

But the possibility of controlling our childrens' genetic destiny strikes me as only good. It is grossly unfair that some families' lives are dominated by the horrors of genetic disease. As a biologist, I know that people suffering from genetic disease are the victims of unlucky throws of the genetic dice. Mutation has been—and always will be—an essential fact of life, because it is through mistakes in gene replication that the positive genetic variants arise which are the lifeblood of evolution. If the gene-copying process were perfect, life as it now exists never would have come about. Genetic disease is the price we pay for the extraordinary evolutionary process that has given rise to the wonders of life on Earth.

Thus, I do not see genetic diseases in any way as an expression of the complex will of any supernatural authority, but rather as random tragedies that we should do everything in our power to prevent. There is, of course, nothing pleasant about terminating the existence of a genetically disabled fetus. But doing so is incomparably more compassionate than allowing an infant to come into the world tragically impaired. There is, of course, the question of who should have the authority to make decisions of this kind. Here the message of past eugenic practices is clear. Never let a government, no matter how apparently benign, into the process. The potential mother should have this authority. It is she who is likely to be most involved with the upbringing of the child.

I am aware that some will argue that the fetus has an inalienable right to life. But the process of evolution never regards any form of life, be it adult or fetal, as an inalienable right. It's better to see humans as wonderful social animals having *needs* (for food, health and sex, for example), *capabilities* (for thought and love, among others) and *responsibilities* (including that to work with other human beings to see that everyone's needs are adequately met). Working intelligently and wisely to see that good genes—not bad ones—dominate as many lives as possible is the truly moral way for us to proceed.

Viewpoint:
All for the Good—Why Genetic
Engineering Must Soldier On

1999

There is lots of zip in DNA-based biology today. With each passing year it incorporates an ever-increasing fraction of the life sciences, ranging from single-cell organisms, like bacteria and yeast, to the complexities of the human brain. All this wonderful biological frenzy was unimaginable when I first entered the world of genetics. In 1948, biology was an all-too-descriptive discipline near the bottom of science's totem pole, with physics at its top. By then Einstein's turn-of-the-century ideas about the interconversion of matter and energy had been transformed into the powers of the atom. If not held in check, the weapons they made possible might well destroy the very fabric of civilized human life. So physicists of the late 1940s were simultaneously revered for making atoms relevant to society and feared for what their toys could do if they were to fall into the hands of evil.

Such ambivalent feelings are now widely held toward biology. The double-helical structure of DNA, initially admired for its intellectual simplicity, today represents to many a double-edged sword that can be used for evil as well as good. No sooner had scientists at Stanford University in 1973 begun rearranging DNA molecules in test tubes (and, equally important, reinserting the novel DNA segments back into living cells) than critics began likening these "recombinant" DNA procedures to the physicist's power to break apart atoms. Might not some of the test-tube-rearranged DNA molecules impart to their host cells disease-causing capacities that, like nuclear weapons, are capable of seriously disrupting human civilization? Soon there were cries from both scientists and nonscientists that such research might best be ruled by stringent regulations, if not laws.

As a result, several years were to pass before the full power of recombinant-DNA technology got into the hands of working scientists, who by then were itching to explore previously unattainable secrets of life. Happily, the proposals to control recombinant-DNA research through legislation never got close to enactment. And when anti-DNA doomsday scenarios failed to materialize, even the modestly restrictive governmental regulations began to wither away. In retrospect, recombinant-DNA may rank as the safest revolutionary technology ever developed. To my knowledge, not one fatality, much less illness, has been caused by a genetically manipulated organism.

The moral I draw from this painful episode is this: Never postpone experiments that have clearly defined future benefits for fear of dangers that can't be quantified. Though at first it may sound uncaring, we can react rationally only to real (as opposed to hypothetical) risks. Yet for several years we postponed important experiments on the genetic basis of cancer, for example, because we took much too seriously spurious arguments that the genes at the root of human cancer might themselves be dangerous to work with.

Though most forms of DNA manipulation are now effectively unregulated, one important potential goal remains blocked. Experiments aimed at learning how to insert functional genetic material into human germ cells—sperm and eggs—remain off limits to most of the world's scientists. No governmental body wants to take responsibility for initiating steps that might help redirect the course of future human evolution. These decisions reflect widespread concerns that we, as humans, may not have the wisdom to modify the most precious of all human treasures—our chromosomal "instruction books." Dare we be entrusted with improving upon the results of the several million years of Darwinian natural selection? Are human germ cells Rubicons that geneticists may never cross?

Unlike many of my peers, I'm reluctant to accept such reasoning, again using the argument that you should never put off doing something useful for fear of evil that may never arrive. The first germ-line gene manipulations are unlikely to be attempted for frivolous reasons. Nor does the state of today's science provide the knowledge that would be needed to generate "superpersons" whose far-ranging talents would make those who are genetically unmodified feel redundant and unwanted. Such creations will remain denizens of science fiction, not the real world, far into the future. When they are finally attempted, germ-line genetic manipulations will probably be done to change a death sentence into a life verdict—by creating children who are resistant to a deadly virus, for example, much the way we can already protect plants from viruses by inserting antiviral DNA segments into their genomes.

If appropriate go-ahead signals come, the first resulting gene-bettered children will in no sense threaten human civilization. They will be seen as special only by those in their immediate circles, and are likely to pass as unnoticed in later life as the now grownup "test-tube baby" Louise Brown does today. If they grow up healthily gene-bettered, more such children will follow, and they and those whose lives are enriched by their existence will rejoice that science has again improved human life. If, however, the added genetic material fails to work, better procedures must be developed before more couples commit their psyches toward such inherently unsettling pathways to producing healthy children.

Moving forward will not be for the faint of heart. But if the next century witnesses failure, let it be because our science is not yet up to the job, not because we don't have the courage to make less random the sometimes most unfair courses of human evolution.

Afterword:
Envoi—DNA, Peace, and Laughter

If, dear reader, you have followed Jim Watson's discourse from his schooldays to his apotheosis as director of one of the world's great laboratories and a grandee—however puckish and irreverent—of the American scientific establishment, you will by now have a passable insight into how modern science functions and be privy also to Watson's robust view of the future.

Watson's early career illustrates vividly the importance of choosing the right academic parentage. German scientists refer felicitously to their research supervisors as their "thesis father" (or perhaps nowadays "thesis mother"): The implication is that they inherit in the course of their apprenticeship a unique intellectual birthright. An earlier Nobel laureate, the biochemist Hans Krebs, once constructed a dynastic genealogy of biochemists who won Nobel Prizes: All without exception sprang from the laboratories of Nobel laureates-to-be. Patronage may of course play its part, but first you have to be bright enough to identify a patron who is not only one of the elect, but whose style will be congenial to you. Watson picked out Salvador Luria, still young and by no means famous, and the course of his life was set.

The quality that above all else marks the great scientist is not skill at solving problems: That comes with the license to practice. The trick is to ask the right questions in the first place—those that need to be answered and can be. There are questions so broad and all-encompassing—how the brain works, how to cure cancer—as to be intellectually vapid. the great nineteenth-century physicist Ludwig Boltzmann put it like this: "The scientist asks not what are currently the most important questions, but

'Which are at present solvable?' or sometimes merely 'In which can we make some small but genuine advance?' [A hundred years on only the modest tone has an alien ring.] As long as the alchemists sought only the philosophers' stone and aimed at finding the art of making gold, all their endeavors were fruitless; it was only when people restricted themselves to seemingly less valuable questions that they created chemistry. Thus natural science seems to lose from sight the large and general questions; but all the more splendid is the success when, groping in the thicket of special questions, we suddenly find a small opening that allows a hitherto undreamt of outlook on the whole."

Watson's testimony explains why cancer research was for so long held in low regard by reputable biologists: Before the startling revelations that emerged from the molecular genetics of bacteriophages, of yeast, and of fruit flies, about how cells work, and why they multiply or die, there was little hope of finding out what goes wrong when a cell becomes malignant. And so, as Watson insists, the monies so lavishly disbursed by the National Cancer Institute achieved little, because ideas were scarcer than research grants. The moral is that the direction of scientific research should not be dictated by the political exigencies. Politicians, both in America and in Europe, have repeatedly displayed a deep misunderstanding of the process of scientific discovery. The tradition stretches far back in history. "But what use is it?" the Chancellor of the Exchequer, William Gladstone, demanded of Faraday after witnessing a demonstration of electromagnetism at the Royal Institution. "I don't know," came the reply, "but one day, sir, you may be able to tax it." It is the very projects (those studies on creatures like frogs and fruit flies) that were excoriated and mocked by Senator Proxmire in the 1970s, with his Golden Fleece awards, that have opened new vistas on genetic disease and cancer. Watson has committed his own institution, the Cold Spring Harbor Laboratory, to the study of cancer because this has at last matured into a subject fit to engage the imagination of bright young (and even not so young) scientists.

The discoveries of molecular biology have already engendered momentous upheavals in clinical practice. There have been remarkable innovations in the treatment and diagnosis of hereditary diseases, of immune conditions, and of a range of cancers, such as childhood leukemias. But increased longevity has brought with it a relentless rise in the incidence of cancer, and politicians and the public have become impatient with the perpetual promise of a consummation that will at least make death optional. Yet scientific medicine (more or less) has been around since the time of Harvey and Paré and their contemporaries (or some would say for the two millenia since Hippocrates of Cos or Galen), while molecular biology is less than 50 years old and has grown into a major enterprise with commercial offshoots only in the last 20. Cancer, the greatest public preoccupation, is not one disease, but a myriad of disorders, caused by malfunctions in cellular control mechanisms. There will be no single "cure," but nobody in the business doubts that the next decade or two will see the realization of many of the hopes for assured detection, prevention, and treatment.

The allure of molecular genetics carries its dangers, of course. It has generated, especially among the more excitable personalities on the wilder fringes of clinical research, an injudicious gold rush. The worst manifestation has been a rash of press releases announcing specious triumphs in the treatment of cancer or of some dire hereditary disease, most of all perhaps in applications of gene therapy (the glamorous prospect of curing a patient by permanently correcting the faulty gene). Expectations have been raised, patients have suffered, and science has been brought into disrepute. (For a magisterial, often shocking, survey of this scene, Stephen Hall's absorbing book, *A Commotion in the Blood,* cannot be bettered.) The spokesmen for the profession must follow where Jim Watson has led, take the public into their confidence, and expatiate with candor, not hyperbole, on the prospects and the problems.

Watson, it will be clear, is an optimist. He scorned the jeremiads that emanated from the opponents of the new biology; they foresaw only ecological catastrophe and a populace ravaged by cancer-bearing bacteria, escaped from the laboratories. So far at least, Watson has been proved right: No one has died of molecular biology, not even any of the researchers who daily manipulate cancer viruses. With increasing knowledge, the fear has receded and, in Watson's view, biological research promises only betterment of the human lot.

Where then will the giddy pace of progress in molecular genetics lead us? As the great physicist Niels Bohr is supposed to have observed, "Prediction is very difficult, especially of the future." Watson is confident, and few would disagree, that the completion of the human genome sequence will have prodigious practical consequences. It will advance diagnosis through identification of mutations in specified genes (indeed an American company already markets a chip bearing 400,000 fragments of DNA, which should in principle allow the detection of a mutation in any one of the human genes); it will lead to the recognition and isolation of gene products—the proteins to which the genes give rise—and it will stimulate the systematic study of the functions (and defects) of these proteins, as targets for a new generation of drugs; it will lead to an understanding of the development of the embryo; it will advance investigations into the relation between genetics and behavior; and it will bring the prospects of gene therapy for hereditary diseases that much nearer. We may also expect that genetic manipulation of skin cells, say, will allow hearts, livers, or kidneys to be grown in the laboratory for spare-part surgery, free from the hazards of rejection. Transgenic pigs—animals with modified or alien genes—are already being bred for organs devoid of the surface antigens that would otherwise provoke instant rejection if implanted into people. And pharmaceutically useful substances will increasingly be harvested from the milk of transgenic farm animals or from the seeds or roots of genetically modified crops; there is even talk of generating in the same way raw materials for industrial chemistry, for the manufacture, for example, of biodegradable plastics.

The genome sequence will also bring with it the social and moral problems to which Watson has alluded. Biological research has become expensive, labor-intensive, and fraught with commercial implications. This will be all the more true of the clinical treatments that will result. To what extent, then, should wealth be permitted to purchase the health and longevity denied to the poor? The specter of eugenics, a social evil rejected and, it seemed, safely buried after World War II, stalks once more in a new guise. Should we try to save babies, or indeed fetuses, whose lives, we can predict, will be blighted by suffering? Or play God by enabling naturally lethal genes to multiply in the population? What value will the taxpayer want to place on the life of a newborn baby? Should information about a citizen's liability to cancer or Huntington's disease be divulged to his insurance company, or the government, or even to himself and his family? And what will be the consequences if the mechanisms by which we age are brought to light and a clever gerontologist discerns the means of arresting them? The litany goes on and the problems can be expected to burgeon. We may also worry that so much of biology is already now concerned with patents, profits, and the Dow Jones. There is no escaping Mark Twain's adage that progress is the exchange of one nuisance for another nuisance.

To most biologists, and to philosophers and politicians also for that matter, manipulation of the germ line—the DNA that passes from parent to offspring—is the last frontier from which mankind will recoil and which must never be crossed. But how are we to prevent an eager geneticist from undertaking such experiments? The molecular biologist Lee Silver, in his somewhat dystopian vision of the the next millenium, *Remaking Eden*,[1] asserts that in a few years from now human cloning (like that of Dolly the sheep) will be feasible (if expensive). A skin cell will suffice for any woman to give birth (herself or through a surrogate mother) to her own baby twin sister, with no messy male intervention. A company registered in the Bahamas already offers—prematurely, to be sure—this service on the Internet for a mere $200,000. Women in their sixties bear children, and the menopause can be reversed. A recent report that a chance ectopic pregnancy was brought to term and resulted in a clutch of healthy triplets gives substance to the prospect of male surrogate mothers, and it will certainly be possible to make a mouse serve as a repository of human sperm. There is plenty more in like vein to tax the imagination.

Yet these are all trifling prospects compared to an assault on the human germ line and the emergence of designer babies, or as a German philosopher has dubbed it, anthropotechnology. There are those who argue that if it is possible it will soon become, to somebody somewhere, irresistibly alluring; so it will inevitably happen, and will indeed in time become as commonplace as the once widely abhorred in-vitro fertilization is now. Couples (or lone mothers) will soon be able to sort through a collection of embryos and select for a place in our midst those with the most desirable gene profiles. And why indeed would those parents in our age of excess, who can afford to press on their offspring all the advantages that health and education can afford, choose to deny them any available genetic privileges at birth? Genes for blond hair, blue eyes, an imposing stature, and perhaps resistance to heart disease should be easy; those that help make a great artist, say, a whole lot less so. Lee Silver in his book

[1]*Remaking Eden: Cloning and Beyond in a Brave New World* by Lee M. Silver (Avon Books, New York, 1997; Weidenfeld and Nicolson, London, 1998).

paints a yet more lurid picture, for he envisages that a genetically enhanced, or as he terms it, "GenRich," population will spring from the affluent stratum of our society and will eventually diverge so far from the proletarian autochthons that the two will no longer be able to interbreed, and it will be a new species of *Homo* that inherits the riches of the earth.

Such apocalyptic fancies are remote from Jim Watson's vision of the biological future. He is concerned with practicalities—with the problems and prospects that will confront us over the next decade or so—and toward those he retains an unquenchable optimism. He does not believe that the scientific community will ever allow the endeavor that it cherishes to run riot. The germ line will remain sacrosanct, to be touched only *in extremis*—if perhaps a threat should arise to the survival of the species, from, say, a new virus. Yet it is undeniable that there has been a new upsurge of public mistrust of science and its works. This, more than an often justified disapproval of the ways of multinational agro-business, has animated the recent campaign against genetically improved crops. Jim Watson has fought such obscurantism with patience, lucidity, and reason. The *enfant terrible* at 70 has lost none of his evangelical enthusiasm for science and its uses; he is still captivated by the beautiful molecule DNA that his work with Crick erected as an icon for our age, and by the richness and promise of the science that has flowed from their discovery.

Walter Gratzer
London, October 1999

About the Authors

James D. Watson

In 1953, while working at the Cavendish Laboratory at Cambridge University, James D. Watson and Francis H.C. Crick determined the double helical structure of DNA. For their discovery, they, with Maurice Wilkins, were awarded the 1962 Nobel Prize in Physiology or Medicine.

Dr. Watson arrived at Harvard University as an assistant professor in 1956 and was promoted to professor in 1961. Soon after, he began a writing career that gave rise to the seminal text, *Molecular Biology of the Gene* (1965), and the autobiographical fragment, *The Double Helix* (1968). The same year, while retaining his position at Harvard, he became director of Cold Spring Harbor Laboratory, and shifted its research focus to the study of cancer. He remained on Harvard's faculty until 1976, when he became the full-time director of Cold Spring Harbor Laboratory. In 1988, Dr. Watson was appointed as associate director of the National Institutes of Health (NIH) to help launch the Human Genome Program. A year later, he became the first director of the National Center for Human Genome Research at the NIH, a position he held until 1992. When Bruce Stillman became director of Cold Spring Harbor Laboratory in 1994, Dr. Watson became its first president.

Dr. Watson was elected to the National Academy of Sciences in 1962, and in 1977, received from then-President Gerald Ford the Medal of Freedom. In 1981, he was elected a member of the United Kingdom's Royal Society. He has received honorary degrees from several universities, including the University of Chicago (1961), Harvard University (1978), the University of Cambridge (1993), and the University of Oxford (1995).

In 1993, the Royal Society awarded him the Copley Medal. In December 1997, President William Clinton awarded him the National Medal of Science for his involvement with the Human Genome Project.

In addition to his *Molecular Biology of the Gene* and *The Double Helix*, Dr. Watson has coauthored *The DNA Story, Molecular Biology of the Cell*, and *Recombinant DNA: A Short Course*.

Dr. Watson was born in Chicago, Illinois, in 1928. He earned a B.S. degree from the University of Chicago in 1947, and a Ph.D. from Indiana University in 1950. In 1968, he married the former Elizabeth Lewis; they have two sons, Rufus (born in 1970) and Duncan (born in 1972).

Walter A. Gratzer

Walter Gratzer is Emeritus Professor of Biophysical Chemistry at London University. He graduated in chemistry from Oxford University and received his Ph.D. at the National Institute for Medical Research in London. He was a Research Fellow at Harvard University. Since then he has pursued his research at King's College, London, most of the time in the Medical Research Council's Cell Biophysics and Muscle and Cell Motility Units. He has been a frequent contributor to *Nature* and edited *A Bedside Nature* and *The Literary Companion to Science*.

Name Index

Amaldi, Edoardo, 37
Amis, Kingsley, 121
Arber, Werner, 39
Avery, O.T., 147–149

Bailey, Kenneth, 18
Baltimore, David, 55–56, 152, 155
Barlow, Horace, 17
Baur, E., 189
Beadle, George, 118
Beckwith, Jonathan, 194
Benzer, Seymour, 149
Berg, Paul, 55
Berget, Susan, 153
Binet, Alfred, 186–187
Bishop, Mike, 155
Bohr, Niels, 9, 13, 212, 233
Boltzmann, Ludwig, 231
Botchan, Mike, 152
Botstein, David, 199
Boyer, Herbert, 55, 61, 113, 153–154, 194
Brachet, J., 23–24
Bragg, Lawrence, 17, 21, 27, 34–36
Brenner, Sydney, 19–20, 26–27, 32–33,
    50–51, 56, 120
Broker, Tom, 153
Bruce, Victor, 14
Burgess, Dick, 51
Burk, Dean, 50
Butenandt, Adolf, 198–199, 217–218

Cairns, John, 52
Califano, J., 68, 70, 78
Caspar, Don, 20
Caspari, Ernst, 10
Chargaff, E., 50–51
Chase, Martha, 40–41
Chibnall, A.C., 18
Chow, Louise, 153
Churchill, Winston, 119
Cleland, Ralph, 125
Clinton, Bill, 44–45, 238
Cohen, Seymour, 10–11, 14, 50, 149
Cohen, Stanley, 55, 61, 113, 154, 194
Collins, Francis, 200
Crawford, Lionel, 53
Crick, Francis, 15, 17–20, 24, 26–28, 32,
    34–35, 39, 41, 43, 50, 97, 119–121,
    125–126, 149–150, 213, 215, 236–237

Darlington, C.D., 19
Darwin, Charles, 4, 111, 175, 181–182, 228
Dausset, Jean, 199
Davenport, Charles B., 183–186, 193, 218
Davis, Ron, 199
Delbrück, Manny, 12, 14
Delbrück, Max, 9–13, 15, 19, 37–40,
    212–213, 218
Delius, Hajo, 53
Demerec, Miloslav, 10–11, 54, 193–194
Doermann, A.H., 12

Donis-Keller, Helen, 199
Doty, Paul, 120
Duesberg, Peter, 155
Dulbecco, Renato, 8–12, 52–53, 151–152, 171
Dunitz, Jack, 12
Dyson, Freeman, 77

Eddy, Bernice, 151
Edwards, R.G., 83–84, 86–87
Einstein, Albert, 215, 227
Eisenhower, D., 44
Erickson, Ray, 155

Faraday, M., 232
Farber, Sidney, 140
Farrell, James, 119
Fermi, Enrico, 37
Fischer, Eugen, 189, 191, 198, 218–219
Fisher, R.A., 18
Fitzgerald, F.S., 117, 121
Folkman, Judah, 158–159
Ford, Gerald, 237
Fraenkel-Conrad, Heinz, 28
Franklin, Rosalind, 120, 125–126
Frederickson, Don, 57, 68, 78
Frisch, Otto, 212

Gajdusek, Carleton, 12
Galen, 233
Galton, Francis, 182
Gamow, George, 25–27
Ganten, Detlev, 215, 218, 220
Gelinas, Rich, 153
Gesteland, Ray, 53, 153
Gierer, Alfred, 28, 149
Gilbert, Walter, 22, 32, 55, 120, 195
Gladstone, William, 232
Goddard, Henry, 186
Green, Howard, 53
Greenberg, Daniel, 136–137
Greene, Graham, 121
Griffiths, John, 27

Gros, Francois, 32
Gross, Ludwik, 151
Grosz, George, 215
Grunberg-Manago, Marianne, 29
Gurdon, John, 83

Hahn, Otto, 212
Haldane, J.B.S., 18
Hall, Ben, 31
Hall, Stephen, 233
Hallervorden, Julius, 221
Hanahan, Douglas, 158
Handler, Phil, 55
Harlow, Ed, 157–158
Harriman, E.H., 183
Harvey, W., 233
Hassel, John, 54
Heisenberg, Werner, 215
Hershey, Alfred Day, 12, 23, 38–39, 40–42
Himmler, 192
Hinshelwood, Cyril, 19, 38
Hippocrates, 233
Hitler, Adolph, 188–191, 197, 208, 217–218
Hoagland, Mahlon, 30, 150
Hodgkin, Allen, 17
Hogness, Dave, 65
Holweck, Fernand, 37
Hoover, J. Edgar, 43
Hopkins, F.G., 17
Huebner, Robert, 153
Hutchins, Robert, 4
Huxley, Andrew, 17
Huxley, Hugh, 17

Isherwood, Christopher, 120

Jacob, Francois, 31–32, 120
Johannsen, Wilhelm, 180
Johnson, Earl, 151
Johnson, Louise, 33
Johnson, Willard, 118

Kalckar, Herman, 13–14
Keilin, David, 18

Keller, Walter, 152
Kendrew, John, 15, 17, 21, 39, 125
Kennedy, J.F., 44
Kennedy, Ted, 66, 68, 140
King, Mary Claire, 203
Kirkwood, John, 12
Klessig, Dan, 153
Kohler, George, 114
Kornberg, Arthur, 51, 54, 149
Kozloff, L.M., 14
Krebs, Hans, 231
Kreisel, George, 126

Laker, Mary, 140
Land, Helmut, 157
Landon, Alf, 118
Lang, Daniel, 120
Lark, Gordon, 14
Lasker, Albert, 147
Lasker, Mary, 147
Laughlin, Harry P., 185–186, 193, 213–214
Lederberg, Joshua, 13–14, 38, 86
Lee, Wen-Hwa, 157
Lenin, 221
Lenz, Fritz, 189, 198, 217
Levine, Arnold, 158
Levinson, Art, 155
Lewis, Elizabeth, 36, 214, 219, 222, 238
Lewis, Herman, 55
Lipkin, David, 29
Lipmann, Fritz, 30, 42
Littlefield, John, 151
Lodish, Harvey, 57
Lukanidin, Eugene, 54
Luria, Salvador E., 8–15, 19, 23, 37–40, 49,
    91, 123–125, 148, 150, 171, 193, 196,
    231
Lwoff, Andre, 28
Lysenko, Trofim, 189

Maaløe, Ole, 14–15
MacLeod C.M., 147–149
Mammen, Jeanne, 215

Marcus, Phil, 52
Markham, Roy, 18, 29
Maxam, Alan, 195
McCarthy, J., 43
McCarty, M., 147–149
McClintock, Barbara, 10
Medawar, Peter, 123
Meitner, Lisa, 212
Mendel, Gregor, 179–183
Mengele, Josef, 198, 217, 219
Meselson, Matt, 32
Michaelis, Leonor, 11
Milstein, Cesar, 114
Mitchell, Peter, 18
Mitchison, Murdoch, 17
Mizutani, Satoshi, 155
Monod, Jacques, 31, 120
Moore, Charles, 214
Mulder, Carel, 53, 152–153
Muller, Hermann J., 7–8, 148, 189
Müller-Hill, Benno, 198, 214

Nathans, Dan, 55, 153
Nixon, Richard, 140
Nomura, Masayasu, 31, 49
North, Tony, 34
Novick, A., 11–12

Ochoa, Severo, 29
Orgel, Leslie, 26–27

Parada, Luis, 157
Paré, 233
Pauling, Linus, 12, 27, 42–45, 117, 126, 137
Pauling, Peter, 34
Payne, Fernandus, 7
Perutz, Max, 15, 17, 21, 125
Phillips, David, 33
Pollack, Bob, 53–55
Porter, Rodney, 18
Proxmire, W., 232
Ptashne, Mark, 41
Putnam, F.W., 14

Rasetti, Franco, 37
Redford, Robert, 71
Ressovsky, Nicolai Timofeeff, 212, 221
Rich, Alex, 25, 27, 29
Richardson, John, 51
Rifkin, Jeremy, 196
Rio, Geo, 37
Risebourgh, Bob, 50
Roberts, Richard, 10, 54, 153
Roosevelt, Franklin Delano, 3, 71, 91, 118
Rosenberg, Alfred, 219
Rosenmann, Sam, 118
Rothschild, Victor, 17
Rubin, Harry, 154
Ruley, Earl, 157
Rutherford, Ernest, 17, 21

Sachs, Leo, 151
Sambrook, Joe, 52, 152–153
Sanger, Fred, 18, 20, 26, 55, 195
Sato, Gordon, 52
Schinkel, Karl Frederick, 214
Schmidt, Benno, 134, 140
Schmidt, Harrison, 68
Schramms, Gerhard, 28
Schrödinger, Erwin, 5, 123, 212
Scolnick, Ed, 157, 199
Shapiro, James, 194
Sharp, Phil, 152–153
Sheldon, W., 10
Sherwood, Robert, 118
Silver, Lee, 235
Singer, Maxine, 55
Smith, Ham, 153
Smith, John, 18
Smith, Kenneth, 18
Soane, John, 214
Soll, Dieter, 55
Spiegelmann, Sol, 31, 49
Stalin, J., 111, 221
Stanley, Wendell, 28
Stent, Gunther, 12, 14–15
Stephenson, M., 30

Steptoe, P.S., 83–84, 86–87
Stewart, Sarah, 151
Stillman, Bruce, 237
Stockman, David, 139
Stoker, Michael, 52, 151
Strassmann, Fritz, 212
Sugden, Bill, 53, 153
Sutton, Walter, 180
Swann, Michael, 17, 19
Sylbalski, Waclaw, 56
Szent-Gyorgyi, Albert, 137
Szilard, Leo, 11–13, 31–32

Tatum, Edward, 38
Tegtymeyer, Peter, 54
Teller, Edward, 44
Temin, Howard, 152, 154–155
Ternan, Lewis, 187
Thomas, Charles, 53
Thomson, J.J., 21
Tissieres, Alfred, 29–31, 34
Tjian, Bob, 55
Todaro, George, 153
Todd, Alex, 18
Travers, Andrew, 51
Tsui, L.C., 200

Van Potter, 49, 150
Varmus, Howard, 155
Vogelstein, Bert, 158–159
Vogt, Oscar, 221
Vogt, Peter, 154
von Verschuer, Otmar, 198–199, 217, 219

Waddington, C.H., 19
Warburg, Otto, 50, 219
Watson, Dudley Crafts, 119
Watson, Duncan, 238
Watson, Elizabeth. *See* Lewis, Elizabeth
Watson, Rufus, 238
Weber, Klaus, 54
Weinberg, Bob, 156–158
Weiss, Paul, 51, 124–125

Weissman, Sherman, 55
Welles, Orson, 119
Westphal, Henry, 152
Wexler, Nancy, 174
White, Ray, 199
Wigler, Mike, 156–157
Willkie, Wendell, 118
Wilkins, Maurice, 34, 120, 237
Williams, Carroll, 52

Williams, Robley, 28
Williams, Shirley, 70
Wilson, Tom, 33, 35
Wright, Sewell, 5, 148
Wulfing, Elke, 216, 218

Yerkes, Robert M., 187

Zamecnik, Paul, 23–24, 30, 150
Zinder, Norton, 51

Subject Index

Abortion, views of Watson, 168, 175–176,
207, 224–225
ACS. *See* American Cancer Society
Adenovirus
E1A role in cancer, 157–158
early research, 153–154
Aloofness, necessity in academia, 109–116
Alzheimer's disease, predisposing genes,
223–224
American Cancer Society (ACS), founda-
tion, 147–148
American scene, science relationship, 105–
108
Animal rights, views of Watson, 166–167
Asimolar, meetings on recombinant DNA
restrictions, 55–57, 63–67

Bacteriophage λ, complementary DNA tail
discovery, 41
Bacteriophage T2
DNA metabolism genes, 50
RNA studies, 31
X-ray inactivation studies, 9, 11–13
Bacteriophage T4
DNA metabolism genes, 50
tryptophan interactions, 12
Behavior, genetic origins, 204–206
Biohazard
recombinant DNA technology
Asimolar meetings, 55–57, 63–67

British regulation, 70, 78
burden of regulations, 76–78
congressional response, 67–68
containment facilities, 58, 75
Department of Health, Education,
and Welfare response, 68–69,
78–79
early concerns, 55
environmentalist response, 67,
71–74
moratorium, 56, 63–65, 195
National Institutes of Health
response, 57–59, 65–66, 68,
72, 75, 77
press response, 56, 58
Recombinant Advisory Committee,
68–70, 78–79
safe strain development, 56–57, 67
tumor virus research, 53–54
Biohazards in Biological Research, publication, 56
Bragg, Lawrence
biography, 34
The Double Helix, Foreword author-
ship, 33–36
BRCA1, cloning and screening, 203–204

Cancer research
American Cancer Society foundation,
147–148

Cancer Research (*continued*)
    angiogenesis, 159
    comprehensive research centers,
        132–134, 136
    cooperation versus competition be-
        tween research groups,
        100–102
    foundations of research, 129–132
    funding
        cuts in the 1980s, 139–140, 142
        requests and expectations, 130–132,
        135–136
        sources in early research, 147–148
    media attacks on progress, 136–137,
        139
    oncogenes, 142–143, 154–159
    prevention by lifestyle modification,
        140
    program prioritization, 141–145
    prospects for cure, 159–160
    tumor viruses, 141–144
    war on cancer legislation, 132–134
CF. *See* Cystic fibrosis
Cloning, humans
    abuse potential, 83, 85
    accessibility of techniques, 87–88
    challenges, 85, 87
    embryo research ethics, 88–89
    fertilization, in vitro, 83–84, 86–87, 229
    germ cell research prohibitions,
        228–229
    international scope of regulation,
        89–90
    media response, 85
    public opinion and involvement,
        86–88, 90
    status, 235
Codon
    assignments, 32
    early speculations, 25–29
Cold Spring Harbor Laboratory
    directorship of Watson, 52–53, 118,
        152, 237

Eugenics Record Office
    closing, 193
    genetic counseling, 184
    immigration views and impacts,
        186, 213
    mental illness attitudes, 184–185
    pedigree assembly, 183–184
    summer visits by Watson as graduate
        student, 10–11, 14, 41
Commercial interests
    evolution of science, 111–113
    genetic engineering, 112–114
    Human Genome Project, 209–210
    monoclonal antibodies, 114–116
Competition. *See* Unpublished informa-
        tion, dissemination
Cystic fibrosis (CF)
    gene mapping, 200
    gene therapy, 170–171

Disillusionment, scientific career, 99–100
Dominant gene disorders, overview,
        182–183
The Double Helix
    Foreword by Sir Lawrence Bragg, 33–36
    writing and publication, 120–121, 237
Down's syndrome, prenatal diagnosis, 170

E1A, role in cancer, 157–158
ELSI. *See* Ethical, societal, and legal issues
Enjoyment, role in scientific success,
        125–126
Ethical, societal, and legal issues (ELSI),
        programs, 173–174, 202–203
Eugenics
    Eugenics Record Office
        closing, 193
        genetic counseling, 184
        immigration views and impacts,
        186, 213
        mental illness attitudes, 184–185
        pedigree assembly, 183–184
    goals, 183

Human Genome Project, arguments
    against, 174–175
Nazi eugenics
    academia response, 197–198
    euthanasia policy of mercy killing,
        191–192, 205
    Final Solution, 192
    Gypsy persecution, 190
    historical background, 188–189
    Jewish persecution, 189–193
    Kaiser Wilhelm Institutes, 218–219
    master race, 189, 208
    modern impact on German re-
        search, 197, 209–212,
        216–218
    record collections, 191
    Russian views, 189
    sterilization laws, 188, 190–191, 214
    tainting of modern research, 208
    war criminals in science, 198–199,
        210, 217, 221
    origins, 182
Evolution
    gene mutations, 163, 181
    overview, 163–164
Excellence
    imitation of role models, 117–119
    versus excellent, 117

Fallback person, role in scientific success,
    125
Funding
    cancer research
        cuts in the 1980s, 139–140, 142
        requests and expectations, 130–132,
            135–136
        sources in early research, 147–148
    expectations of science by public,
        106–107
    politicians as source, 109–110

Gene therapy
    cystic fibrosis, 170–171

prospects, 206
Genetic engineering. See Recombinant
    DNA
Genetic screening
    abortion, views of Watson, 168,
        175–176, 207, 224–225
    availability, 177
Germ cell research, prohibitions, 228–229,
    235
Gypsy, persecution by Nazis, 190

Hershey, Alfred Day
    biography, 40–42
    Hershey–Chase experiment, 40–41
HGP. See Human Genome Project
Human cloning. See Cloning, humans
Human Genome Project (HGP)
    applications, 171–172
    commercial interests, 209–210
    costs, 200–201
    ethical, societal, and legal issues pro-
        grams, 173–174, 202–203
    genome size and mapping, 171
    heterogeneity of gene pool, 163–164,
        179, 199
    initiation and objections, 171–173,
        200–201
    overview, 169
    potential misuse, 169
    privacy issues, 174
    public reaction, 201
Human rights, views of Watson, 175
Huntington's disease, gene mapping, 200,
    203, 223

Intelligence, role in scientific success, 124
Intelligence quotient (IQ), testing and
    early justification of racial
    discrimination, 186–188
IQ. See Intelligence quotient

Jews, persecution by Nazis, 189–193

Language, importance to civilization, 165

Luck, role in scientific success, 123–124
Luria, Salvador E.
  biography, 37–40
  Indiana University graduate school
    experience of Watson, 7–14,
    23, 231

Mendelian heredity, overview, 179–180
Mental illness
  eugenics views, 184–185
  genetic origins, 204–206
  socialist views, 205
Messenger RNA, discovery, 31–32, 50–51,
  120
Molecular Biology of the Gene, publication,
  52, 120, 152, 237
Monoclonal antibody
  commercial interest, 114–116
  overview of history, 114–115

Nazi. *See* Eugenics
Networking, role in scientific success,
  125–126

Oncogenes, early research, 142–143,
  154–159

Pauling, Linus, biography, 42–45
Peers, ranking, 93–94
Phage transduction, discovery, 194
Politics
  cancer research
    comprehensive research centers,
      132–134, 136
    funding
      cuts in the 1980s, 139–140, 142
      requests and expectations, 130–
        132, 135–136
    media attacks on progress, 136–137,
      139
  funding, politicians as source, 109–110
  Genes and Politics talk in Berlin, 209,
    212–222
  human cloning

media response, 85
  public opinion and involvement,
    86–88, 90
recombinant DNA technology
  Asimolar meetings, 55–57, 63–67
  British regulation, 70, 78
  burden of regulations, 76–78
  congressional response, 67–68
  Department of Health, Education,
    and Welfare response, 68–69,
    78–79
  early concerns, 55
  environmentalist response, 67,
    71–74
  moratorium, 56, 63–65, 195
  National Institutes of Health re-
    sponse, 57–59, 65–66, 68, 72,
    75, 77
  press response, 56, 58
  Recombinant Advisory Committee,
    68–70, 78–79
  removing from genetic decisions,
    206–208, 232
Publication
  crediting another's work in a manu-
    script, 96
  delay between submission and publica-
    tion, 95
  preprint distribution, 95–96

Recombinant DNA
  biohazards
    Asimolar meetings, 55–57, 63–67
    British regulation, 70, 78
    burden of regulations, 76–78
    congressional response, 67–68
    containment facilities, 58, 75
    Department of Health, Education,
      and Welfare response, 68–69,
      78–79
    early concerns, 55
    environmentalist response, 67, 71–
      74

moratorium, 56, 63–65, 195
National Institutes of Health response, 57–59, 65–66, 68, 72, 75, 77
press response, 56, 58
Recombinant Advisory Committee, 68–70, 78–79
safe strain development, 56–57, 67
commercial interest in genetic engineering, 112–114
development, 61
ethics, 167–168
leftist opposition, 71–73, 196
National Academy of Science review, 62
overview of history, 113
plant engineering, 167
potential, 62, 71–72, 233–234
safety, 195, 228, 233
Restriction enzyme
early research, 153
gene polymorphism analysis, 199
Retrovirus, early research, 155
Ribosome, structure determination attempts, 29–31
Risk-taking, role in scientific success, 124–125
RNA
adaptor hypothesis and transfer RNA, 27, 30, 150
coding hypotheses, 25–29, 32
early speculations, 23–24
genetic material transmission, 149
infectivity, 28
messenger RNA discovery, 31–32, 50–51, 120
structure determination attempts, 20, 25, 29, 119
types, 24
RNA polymerase, early purification, 51
RNA splicing, discovery, 153–154
Rous sarcoma virus (RSV), early research, 155

RSV. See Rous sarcoma virus

Schizophrenia, predisposing genes, 205
Scientific progress, evolution, 110–111
Secrecy. See Unpublished information, dissemination
Social Darwinism, overview, 181–182
Succeeding in science, rules of thumb, 123–126
SV40
biohazards, 53–55
culture and yield, 54–55
oncogene studies, 152–153, 158–159
RNA splicing, 154

TBSV. See Tomato bushy stunt virus
TMV. See Tobacco mosaic virus
Tobacco mosaic virus (TMV), reconstitution, 27–28
Tomato bushy stunt virus (TBSV), RNA coding studies, 29
Transfection
development, 156
oncogenes, 156–157
Transfer RNA, adaptor hypothesis, 27, 30, 150
Transgenic animal, organ harvesting, 234
Tryptophan, interactions with bacteriophage T4, 12
Tumor viruses
biohazards, 53–55
cancer research priority, 141–144
oncogenes in DNA versus RNA viruses, 156
overview of early studies, 149–152, 154–155
RNA splicing, 154

Unpublished information, dissemination
applied human projects, 100
avoidance of certain scientists, 96–97
cancer research, cooperation versus competition, 100–102

Unpublished information, dissemination
(*continued*)
competition, guidelines for handling,
98–99
cooperation versus group size, 102
crediting another's work in a manuscript, 96
delay between submission and publication, 95
friendships with competitors, 103
peer ranking, 93–94
philosophical underpinnings of Watson, 91–92
preoccupation with priority, 93
preprint distribution, 95–96
reactions to being scooped, 94–95
secrecy guidelines, 97–98, 102

Watson, James D.
Berlin experience, 209–222
Caltech experience, 20, 25
Cambridge Cavendish Laboratory
experiences, 17–22
childhood in Chicago, 3–4, 91, 117–119

Cold Spring Harbor Laboratory directorship, 52–53, 118, 152, 237
Copenhagen post-doctoral experience,
13–15
DNA structure elucidation, 215–216,
237
embryology interest, 51–52
Harvard teaching experience, 30, 237
honors, 237–238
Indiana University graduate school
experience, 7–14, 23, 231
marriage and children, 238
optimism, 231–236
political influences, 3, 71, 118–119
publications, 237–238
religious beliefs, 118, 175–176
University of Chicago undergraduate
experience, 4–5, 7
writing aspirations, 120–121
What is Life?, influence on Watson, 9, 120,
123, 212

X-ray inactivation studies, bacteriophage
T2, 9, 11–13